THE PRESIDIO

NORTHEASTERN UNIVERSITY 1898–1998

THE

LISA M. BENTON

PRESIDIO

FROM ARMY POST TO NATIONAL PARK

Northeastern University Press 1998

Copyright 1998 by Lisa M. Benton

All rights reserved. Except for the quotation of short passages for the purposes of criticism and review, no part of this book may be reproduced in any form or by any means, electronic or mechanical, including photocopying, recording, or any information storage and retrieval system now known or to be invented, without written permission of the publisher.

Library of Congress Cataloging-in-Publication Data

Benton, Lisa M.
    The Presidio : from Army post to national park / Lisa M. Benton.
      p. cm.
    Includes bibliographical references (p.   ) and index.
    ISBN 1-55553-335-3 (alk. paper)
     1. Presidio of San Francisco (Calif.)—Planning. 2. Presidio of
San Francisco (Calif.)—History. 3. Presidio of San Francisco
(Calif.)—Geography. 4. San Francisco Bay Area (Calif.)—Geography.
5. San Francisco (Calif.)—Buildings, structures, etc. I. Title.
F869.S38P742   1998
979.4'61—dc21   97-51930

Designed by Janice Wheeler

Composed in Dante by G & S Typesetters, Inc., in Austin, Texas. Printed and bound by The Maple Press Company in York, Pennsylvania. The paper is Sebago Antique Cream, an acid-free sheet.

MANUFACTURED IN THE UNITED STATES OF AMERICA

02 01 00 99 98   5 4 3 2 1

*Dedicated to my mother, Bonnie Jean Okey Benton,
and my father, Robert Bruce Benton, with much love.*

And for John Rennie Short, with the stars in his eyes.

Contents

List of Tables   ix
List of Figures   xi
List of Key Organizations and Individuals   xiii
Acknowledgments   xv

CHAPTER 1
From Post to Park: An Introduction   3

CHAPTER 2
The Presidio, 1776–1846   6

CHAPTER 3
The Presidio as U.S. Army Post, 1846–1906   20

CHAPTER 4
The Presidio in the Twentieth Century, 1906–1989   40

CHAPTER 5
Post–Cold War Military Base Closures: From Swords to Plowshares   67

CHAPTER 6
Planning the Presidio's Conversion: From Post to Park   86

CHAPTER 7
Congress and the Presidio: The Politics of Parks   113

CHAPTER 8
Nature, Culture, and the National Parks   147

CHAPTER 9
Transforming the Presidio: An Analysis and Some Conclusions   178

Appendix A   A Note About the Research for This Project   203
Appendix B   Presidio Council Members, 1993   209
Appendix C   National Park Service Planning and Transition Team Members   213
Notes   215
References   255
Index   271

TABLES

CHAPTER 5
Table 5.1   Base Closure and Realignment Costs and Savings   84

CHAPTER 6
Table 6.1   The Presidio Planning Guidelines   96
Table 6.2   Presidio Park Program Use Concepts   99
Table 6.3   Estimated Costs of the Presidio's Grand Vision   107
Table 6.4   Predicted Costs and Economic Impacts of All Alternative Plans   110

CHAPTER 7
Table 7.1   Presidio Management: A Public-Private Partnership   122
Table 7.2   Presidio Cost Analysis   134
Table 7.3   Measuring Merit?   146

CHAPTER 8
Table 8.1   Units of the National Park System   150
Table 8.2   Public Recreation Areas, 1965   159

Figures

Chapter 1

Figure 1.1   Map: The Presidio in Its Urban Context   2

Chapter 2

Figure 2.1   "A Most Delightful View"   10
Figure 2.2   Map: The Geographical Relationship of Presidio, Pueblo, and Mission   12
Figure 2.3   The Presidio, circa 1816   16

Chapter 3

Figure 3.1   Map: Proposed Presidio Reservation of 1850   23
Figure 3.2   Map: Defending the Golden Gate   24
Figure 3.3   Map: Fort Point at the Presidio in 1860   25
Figure 3.4   Officers' Quarters on Funston Avenue   27
Figure 3.5   Presidio Tree Planting, circa 1887   34
Figure 3.6   The Presidio Wall   35
Figure 3.7   The Presidio Gate, circa 1906   36

Chapter 4

Figure 4.1   Presidio Troops Aid the City, 1906   41

Figure 4.2   Map: The Presidio Grows   43
Figure 4.3   Mission Revival Style at Fort Winfield Scott   44
Figure 4.4   Crissy Airfield, circa 1933   46
Figure 4.5   Construction of the Golden Gate Bridge, circa 1935   49
Figure 4.6   Map: The Golden Gate National Recreation Area, 1990   58

Chapter 5

Figure 5.1   Cycles of Real Defense Spending and Reduction, 1940–1994   68
Figure 5.2   Map: The Gunbelt   71
Figure 5.3   Map: Defense Spending as a Percentage of State Purchases, 1991   72
Figure 5.4   Map: Cumulative Major Base Closures, 1988–1995   85

Chapter 6

Figure 6.1   The Presidio Planning Process: A Timeline of Key Events and Decisions   97
Figure 6.2   Map: The Thirteen Planning Areas of the Presidio   101

Chapter 7

Figure 7.1   Timeline for Implementing the Presidio Plan   114
Figure 7.2   Map: The Presidio According to Duncan   132
Figure 7.3   From Post to Park: A Gala Celebration   139
Figure 7.4   Timeline of the Presidio Trust Legislation   144

Chapter 8

Figure 8.1   The Golden Gate National Parks Logo   148
Figure 8.2   A Hierarchy of National Parks   164

Key Organizations and Individuals

Golden Gate National Recreation Area (GGNRA), established in 1972

People for a Golden Gate National Recreation Area (PFGGRNA)

Golden Gate National Park Association (GGNPA), a nonprofit Park Service partner organization

Bruce Babbitt
Secretary of the Interior (1993–present)

Roger Kennedy
Director, National Park Service (1993–1997)

James Ridenour
Director, National Park Service (1989–1993)

William K. Reilly
Former Administrator, U.S. Environmental Protection Agency
Senior Adviser to the Presidio Council

Representative Bruce Vento
Democrat, Minnesota
Chair, House Subcommittee on National Parks

Representative Nancy Pelosi
Democrat, San Francisco

Senator Dianne Feinstein
Democrat, California

Senator Barbara Boxer
Democrat, California

Brian O'Neill
Superintendent, Golden Gate National Recreation Area (National Park Service)

Robert Chandler
Presidio General Manager (National Park Service)

Don Neubacher
Captain, Presidio Planning and Transition Team (National Park Service)

James Harvey
Chair, Presidio Council

Toby Rosenblatt
Chair, Golden Gate National Park Association Board of Directors

Greg Moore
Executive Director, Golden Gate National Park Association

Craig Middleton
Director for Government Relations, Golden Gate National Park Association

Acknowledgments

Several years ago on a very cold February morning in Syracuse, I was sitting at my desk writing something incoherent about national parks and the Presidio, when the doorbell rang. I opened it to find a man in overalls holding a very large white cardboard box. He handed me the box, said everything was paid for, and told me to have a nice day. I warily opened the box (my brother has hidden rubber snakes in the most imaginative places). Inside were two dozen donuts. I ate most of them and, refueled, returned to my desk and wrote a pretty good chapter. I never found out who sent those donuts, but I have my suspicions.

This is my way of beginning my acknowledgment and thank-yous to family and friends. This way, they know how important they are because they don't have to search through a long list of unfamiliar names, only to find themselves listed in the last paragraph. So first, I must thank my parents, Bonnie Benton and Robert Benton, for their love and support. Without their encouragement, sacrifice, and generosity, I never would have attended Stanford University for my B.A., which means I never would have ended up at Syracuse University doing graduate work, which means I never would have been writing this book when the box of two dozen donuts arrived. Without the love and support and encouragement from John Rennie Short, the love of my life, I never would have thought I was capable of writing a book, so I would not have tried, and who knows if the mysterious sender would have thought I needed donuts. After all, family and friends know a bit of refined sugar in the morning helps my creativity.

There are many other special people who deserve my appreciation (and who are on my list of suspects). Michele Antoinette Judd, my best girl-friend, visited me frequently when others feared to tread in central New York, and when she couldn't visit, she phoned. The Hoover House Staff-less bunch—Jeanne Kennedy, Christine Beckman, Trish Benson, Rose Guntly, and Margaret Dilg—gave me support, advice, comfort, and, frequently, a warm bed and many, many good bottles of wine. Don and Robin Kennedy gave me valuable insight into this project and wise words, and also provided many, many good bottles of wine (see a pattern here?). All of these wonderful people are suspects in the Great Mysterious Donut Caper. Maybe now, when they see their names in print, one of them will confess.

But, you are asking yourself, how did I get started on a book about the Presidio? It all began in the summer of 1993. I sat at a table in Hobee's restaurant in Palo Alto, having lunch with Joan Parker, a friend from my days at Stanford University. She knew I was in the Bay Area to spend my vacation hunting down a dissertation topic. She mentioned that Stanford's Institute for International Studies needed an assistant of sorts: did I need a temporary job? Thus I stumbled my way into fieldwork for this book. So, Parker, thanks for that, among other things.

As it turned out, I worked for Bill Reilly, who hired me for the position by default—he forgot to show up for my interview. But he made up for this inauspicious beginning by letting me accompany him to his many meetings as senior adviser to the Presidio Council. And so I ended up spending six months in San Francisco at the most important time for the Park Service and the Presidio. Without his involvement, I wouldn't have learned about the project. I was very fortunate that I became engrossed with the Presidio project when I did; I also was very fortunate that so many people and organizations were enthusiastically supportive, generous, and very accessible. Robert Chandler, Brian Huse, Judy Lemon, Craig Middleton, Greg Moore, Don Neubacher, and Toby Rosenblatt spent valuable time to talk with me at length about the Presidio's ins-and-outs. I also want to thank the Presidio Council, the Golden Gate National Parks Association, the staff at the Golden Gate National Recreation Area, and the National Park Service for their time and assistance and for access to their documents and materials. Susan Haley and Mary Gentry at the GGNRA Park Archives and Records Center were of great help in providing the historic photographs of San Francisco and the Presidio. Stanford's Institute of International Studies gave me a desk, a phone, a computer, and (unbeknownst to them) lots of free photocopying. More important, the office held the correct attitude toward life: margaritas on Fridays after 5 P.M.

Of course, it isn't enough to obtain the information. You've got to make sense of it. In this regard, I owe a great deal of thanks and appreciation to Professors John Mercer, Mark Monmonier, Jim Newman, and Anne Mosher of Syracuse University. Each of them has taught me something important about words and writing. Thanks also to Mike Kirchoff at the SU cartography lab who patiently assisted me with the maps and illustrations. I also want to acknowledge the generosity of Syracuse University, which provided two years of fellowship money for me to work full-time on this research project.

This was a great project that continued to fascinate me throughout. What follows is my interpretation of events and outcomes, and so any inaccuracies are mine alone.

THE PRESIDIO

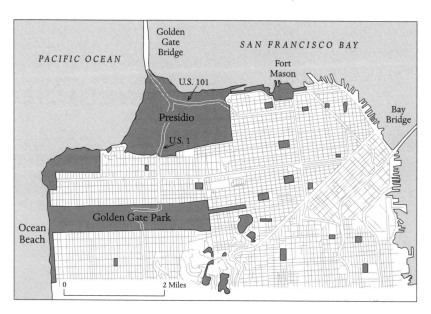

Figure 1.1
The Presidio in Its Urban Context. *Source: Map by author, based on a National Park Service map.*

CHAPTER 1

FROM POST TO PARK: AN INTRODUCTION

This is a chronicle of America's newest national park, the Presidio—how it came to be and how it almost didn't.

Within this chronicle is a collection of subplots: the origins of the Presidio; its evolution as an Army post; its relationship with the city of San Francisco; grass-roots community efforts to protect the post; public participation in planning for the Presidio's conversion and the creation of the Park Service's "Grand Vision"; the battle for congressional approval of the management plan; and the Presidio's context within the national park system.

First, an introduction to the main character—the Presidio of San Francisco.

THE PRESIDIO: AN INTRODUCTION

Nearly 3 square miles in area and twice the size of Central Park in New York, the Presidio is nestled between the city of San Francisco and the Pacific Ocean (see Figure 1.1). From 1776 until 1994, the Presidio guarded the Golden Gate as an Army post under Spanish, then Mexican, then American flags. Within its 1,500 acres is the unparalleled collection of military history, cultural landscapes, recreation areas, and natural features that resulted in its designation as a national historic landmark in 1963.

The Presidio is a park within a city. Monterey cypress and pines guard the ridge line from the strong Pacific winds, and golden grass and Manzanita hug the side of the hills. The Presidio Forest spans 600 acres and has hiking trails, vistas, beachfront promenades, enclaves of wildflowers, and grassy open spaces. All of these features are but a stone's throw from the Golden Gate Bridge, tourist destinations such as Fisherman's Wharf and Ghirardelli Square, several prestigious shopping districts, and exclusive residential neighborhoods.

The Presidio is also a city within a city. The post has more than eight hundred buildings, including an officers' club, a bowling alley, a commissary, a research hospital, warehouses, residences, a movie theater, horse stables, a museum, and a golf course.

This is the complex and diverse place that became the newest national park. But this wasn't always so.

A Biography of Place

The story of the Presidio is a biography of place. Simply defined, a biography is a written account of a life. In this case, the life is that of a place. My intent in this book is to explore why the Presidio has been such a special place to so many people; to document the ways in which people mobilized to protect, define, and give new meaning to the post; and to uncover the reasons underlying the resistance and hostility to the Grand Vision for the newest national park.

The biography begins with a historical geography of the post. It details the establishment of the Spanish *presidio,* its formative years, and its maturation, and the biography documents the ways in which influential people and events shaped its identity, its purpose, and its appearance up until 1989. The historical geography of the Presidio is tied to two broad themes: first, its history as a military post; second, its history as an integral part of San Francisco. The historical geography also provides the context for understanding the processes by which the Presidio's future as an urban national park was ensured.

After the historical geography, the biography documents the effect of broad changes in U.S. defense policy and in the military structure on the Presidio. In 1989, the Defense Department decided the Presidio was obsolete. It was to be closed and transferred to the National Park Service. Thus began a conversion and transformation process that lasted from 1989 to 1997. After several years of planning, in 1993 the Park Service released its Grand Vision for the Presidio. The plan described a global center whose mission would be to address the world's most critical environmental, social, and cultural challenges. It envisioned a new type of national park, one that combined traditional park recreation and solitude with activities and programs sponsored by national and international organizations devoted to improving human and natural environments.[1]

The city of San Francisco reacted to the plan in an unusual way—by enthusiastically endorsing it. The Presidio plan was endorsed by San Francisco's political elites, the business community, the major daily newspapers, and more than thirty nonprofit organizations and neighborhood coalitions. But despite local approval and endorsement, the Presidio plan faced a difficult challenge: securing approval from Congress.

The Presidio plan became embroiled in politics. It often faced hostile and impassioned opposition. This opposition placed the proposed legislation in a three-year stalemate, between 1993 and 1996. During this time, the Presidio plan was debated, negotiated, and compromised. Securing congressional approval of the Presidio plan proved to be a political problem—a contest for the control of space and the future of a national park. This part of the biography is a snapshot of a geographical moment, a moment of transformation brought about by global and national changes shaping events in a particular place.

I use the expression *biography of place* as a way to expand traditional notions of historical geography. Of course places are inanimate objects, but underlying my use of *biography of place* is my belief that there are places that profoundly affect people's lives, places that are infused with meaning and importance. Thus this biography is at once a historical geography, an analysis of contemporary events, and speculation about the future.[2] I aim to illuminate connections between humans and their environment, between historical issues and contemporary issues, between a city and a military park/urban open space. Central to this book is the concept of transformation of place-meaning. Understanding the meaning of a place and the impact of that place on people's lives is the primary purpose of any biography of place. What follows is a biography of the Presidio that represents my interpretation of a unique set of circumstances and opportunities; it is just one of many stories that could be written.[3]

CHAPTER 2

THE PRESIDIO, 1776–1846

THE EARLY YEARS

In 1768, the area around the present-day Presidio of San Francisco was probably home to a group of Ohlones who lived along the shore of the marsh, where the surf batters the continent's edge. Before the coming of the Spaniards, the Ohlone Indians populated the Bay Area, living in thirty or forty permanent villages along the shores of San Francisco Bay.[1] Over 10,000 people lived along the Pacific coastal area stretching from San Francisco Bay to Monterey Bay. Although these native Indians are known as Ohlones, they were far from being a homogeneous population. They spoke from eight to twelve different languages, and although groups of some 250 people were loosely affiliated, there was not a larger tribal organization.[2]

Like many other California Indians, the Ohlones were hunters and gatherers, subsisting in the San Francisco area partly on shellfish taken from the bay. They fashioned tule rush or split redwood bark into lightweight, buoyant boats from which they set out their fishing nets. They gathered the ubiquitous acorns, available from several species of live oak, for bread and paste. Women and children collected strawberries, manzanita berries, wild carrots, onions and herbs, and even grasshoppers. They caught various birds and other small animals with nets. The men hunted deer and elk. They had an intimate knowledge of the animals and plants around them. The Ohlones honored the plants and animals of the landscape, worshipped animals spirits as gods, and felt the power of animal spirits in their dreams. The small area in which the Ohlones lived was a land of many resources, of seemingly inexhaustible plenty, "and for century after century the people went about their daily life secure in the knowledge that they lived in a generous land."[3]

Exactly when the Ohlones moved into the San Francisco Bay area is unknown, but archaeological evidence suggests they may have settled there 5,000 years ago.[4] It is likely that they were among the first peoples to populate and settle along the western coastal bluff that borders the Pacific Ocean and San Francisco Bay—a place now called the Presidio. Anthropologist Alfred Kroeber suggests that the Ohlones achieved relative peace and a stability that lasted for hundreds of years.[5] It would appear they rarely engaged in warfare.

The first human occupancy of the San Francisco Bay area—in particular the present-day Presidio—lasted thousands of years. During this time, the Presidio was a place of community, stability, and peace.

THE SPANIARDS ARRIVE, 1769–1776

In the late 1760s, the Spanish Crown took steps to colonize the coast of California, one of the last regions to come under Spanish influence. Despite explorations in the sixteenth century—notably by Englishman Francis Drake in 1579 by sea, and by Spaniard Juan Cabrillo in 1542 by land—Spain had not established settlements in Alta California.[6] Many factors influenced the Spanish Crown's decision to occupy Alta California in 1769, but perhaps the two most important were political and religious. José de Galvez, Spain's visitador-general in New Spain, sought to increase the Crown's revenues and to fortify the defense against the British and Russians, who were exploring northern California as a possible location for outposts for the booming fur trade.[7] In addition, by the late 1760s, the Franciscan Order, led by Father Junipero Serra, was eager to begin the spiritual conquest of the indigenous peoples of Alta California. Serra seemed driven by a consuming passion for saving souls: "Missions, my lord, missions—that is what this country needs," he wrote to the Spanish commandant general.[8] For these reasons, de Galvez ordered settlements to be established at San Diego and Monterey and assigned Don Gaspar de Portolá to lead the expedition.

In May 1769, Portolá, Father Serra, and sixty soldiers, priests, and servants made their way north from Mexico toward San Diego. They reached San Diego in late June. From there, the group continued on toward Monterey Bay, which Spanish explorer Sebastian Vizcaíno had discovered and described in 1602. Vizcaíno had detailed the anchorage and environs of Monterey as seen from his ship as it sat at anchor in the bay. Portolá and his entourage, however, failed to recognize the landmarks and continued north, camping at the base of a mountain a few miles south of San Francisco Bay. In the morning, Portolá sent Sergeant José Francisco de Ortega and a scouting group north. The November air was crisp and clear, and they easily spied the entrance to the Golden Gate and the white cliffs beyond. Because of the fog bank that usually obscures the view of the harbor from the Pacific, it is likely they were among the first Europeans to look on the waters of the bay and to "discover" the immense prize, the greatest natural harbor on the Pacific coast.[9]

Portolá realized he had overshot Monterey and kept careful accounts of this new discovery. He remembered that José de Galvez had promised Father Junipero Serra that if the explorers discovered a new harbor, it would be dedicated to Francis of Assisi. The scout-chronicler called the harbor Gran Puerto de San Francisco—Bay of San Francisco. It seems that the significance of this discovery escaped him at the time, for Portolá was single-mindedly intent on finding Monterey.[10] Portolá's party returned south to search for Monterey, intending to return the next spring for more detailed scouting.

At the outset of 1775, the Spanish Crown decided to establish a permanent outpost by the Bay of San Francisco and began to make preparations for the planting of Spanish settlements. Early in the year, Commander Juan Manuel de Ayala and his supply ship, *San Carlos,* anchored in the bay. Ayala and his men spent two months making a thorough reconnaissance of the harbor and writing their observations.[11] Ayala reported that the harbor possessed "a beautiful fitness and it has no lack of good drinking water and plenty of firewood and ballast . . . [the Bay] had a healthful climate and docile natives."[12] The area now officially called San Francisco Bay stood ready for European colonization.

In October 1775, Captain Juan Bautista de Anza, along with 250 soldiers and colonists, began an overland journey from northern Mexico to the San Francisco Peninsula. They reached Monterey five months later, in March 1776. De Anza took a small advance party from Monterey north to San Francisco to secure the new outpost of the Spanish Empire.

"A Most Delightful View"

On the eastern, interior side of the peninsula, at its tip, the scouting party chose a site for a mission to be dedicated to Saint Francis of Assisi. The scouts then headed west, toward the Pacific. At the day's end, the party camped on the bank of Mountain Lake, now on the southern edge of the San Francisco Presidio.[13] The next day, de Anza and his soldiers explored the windward northwesternmost tip of San Francisco Peninsula. Looking down from cliffs that afforded a view of both the Pacific Ocean and the harbor entrance, de Anza made his decision: the presidio would be located here. One of de Anza's officers, the diarist Father Pedro Font, wrote down his impressions: "This place and its vicinity has abundant pasturage, plenty of firewood, and fine water, all good advantages for establishing here the presidio or fort which is planned . . . The port of San Francisco . . . is a marvel of nature, and might well be called a harbor of harbors."[14] On March 28, 1776, de Anza erected a cross to mark the site for the presidio, an action that marked official possession of this land by Spain.

The primary function of the Presidio of San Francisco was to mark, protect, and preserve Spain's newest possession. Father Font wrote the following description in his diary:

> This mesa the commander selected as the site of the new settlement and fort which were to be established on this harbor; for, being on a height, it is so commanding that with muskets it can defend the entrance to the mouth of the harbor, while a gunshot away it has water to supply the people, namely the spring or lake where we halted. The mesa is very open, of considerable extent, and level, sloping a little toward the harbor. It must be about half a league wide and somewhat longer, getting narrower until it ends right at the white cliff. *This mesa affords a most delightful view* [see Figure 2.1], for from it one sees a large part of the port and its islands, as far as the other side, the mouth of the harbor, and of the sea all that the sight can take in as far as beyond the farallones. Indeed, although in my travels I saw very good sites and beautiful country, I saw none which pleased me so much as this. And I think that if it could be well settled like Europe, there would not be anything more beautiful in all the world, for it has the best advantages for founding in it a most beautiful city, with all the conveniences desired, by land as well as sea, with that harbor so remarkable and so spacious, in which may be established shipyards, docks and anything that might be wished [my emphasis].[15]

Father Font entered this description in his diary on the day of the founding of the Presidio of San Francisco, Spain's fourth presidio in California, following those at San Diego, Santa Barbara, and Monterey.

Pueblos, Missions, and Presidios

In 1776, Spain's approach to colonization in the Americas centered on the establishment of three interdependent institutions—pueblos, missions, and

Figure 2.1
"A Most Delightful View." This view of the Pacific Ocean and bay entrance from the coastal bluffs at the Presidio may be the site that inspired Father Pedro Font's diary entry in 1776. *Source: Photograph by author.*

presidios—to ensure the secular, spiritual, and military conquest of the indigenous peoples. Although these institutions were interdependent, each had its own governing structure. The pueblos were commanded by a representative governor sent by the Spanish Crown; the missions were under the jurisdiction of a "Father-President"; the presidios were under the command of an army officer.

The pueblo, a civilian settlement, had as its main purposes the provision of food for the region and the creation of a stable Spanish population base. Settlers included traders, miners, and farmers, all of whom were expected to create a civilian economy.

The mission, designed for bringing the Indians into the "True Faith," had as its primary function the transformation of indigenous peoples into a Catholic labor force that could be employed eventually in farming and ranching. Once established, the mission was to become a self-supporting economic unit that would enhance the security of the territory. In theory, missions were intended to be transitional institutions. Once the Indians had been converted to Catholicism, had been taught to pray properly, eat with spoons, and wear clothes, and had learned farming, the missions were to be dismantled, leaving a stable, devoted secular order.[16]

The presidio, or military garrison, was the defensive arm of Spanish settlement and was the location of both a military and a civilian complex.[17] Troops were essential to the conquering and settlement of California. Presidio troops also supported the work of the padres and protected the missions. As National Park Service historians Langelier and Rosen explain, "derived from the Latin term *presidium,* a fortified or garrisoned place, the Spanish presidio acted as the advance guard of territorial settlement. In addition to its military role, it provided the core of government, social, and economic activity in the region."[18]

Eventually, in Alta California, twenty-one missions were established along the coast, each separated by a day's journey of some 40 miles along a route the Spanish called El Camino Real. There were almost as many pueblos but only four presidios in all of Alta California: at San Diego, Santa Barbara, Monterey, and San Francisco. Though only four in number, the presidios held thousands of Indians in check and were symbols of Spanish power, deterring foreign incursions.[19] Spain's plan for the conquest and settlement of Alta California, and specifically for the San Francisco Bay harbor, depended on the establishment of these three interdependent institutions (Figure 2.2). Since the ultimate goal of presidio-mission-pueblo settlement was to ensure Spain's claim to territory, the choice of settlement sites was important. The site chosen for the presidio was the most important, for it needed to provide maximum advantage against foreign invasion and hostile Indian attacks; the siting of pueblos and missions could be more flexible.[20]

Although presidios played essentially the same role as French and British forts, the Spanish presidios in Alta California faced additional obstacles. First, they were challenged to maintain a very long frontier line, a frontier line for which Spain provided only limited financial support. In the first five years of occupation, there were only sixty-one soldiers in all of Alta California.[21] They were responsible for the safety of the missions and pueblos along the 400-mile coastline. Second, the presidios differed from the French and British forts in the degree to which they were controlled by the mother country. Spain discouraged self-reliance; the Spanish government carefully controlled every detail of life in the presidios through numerous regulations and constant supervision. This strict regimentation crippled the capacity of the people to adapt to new, unpredictable conditions.[22] Also, the presidio faced problems associated with unreliable sources of supply—shortages of food and weapons were commonplace. As a result, from the outset, confusion of both purpose and priority characterized the Spanish settlement of San Francisco, a fateful duality from which the Presidio would never fully recover.

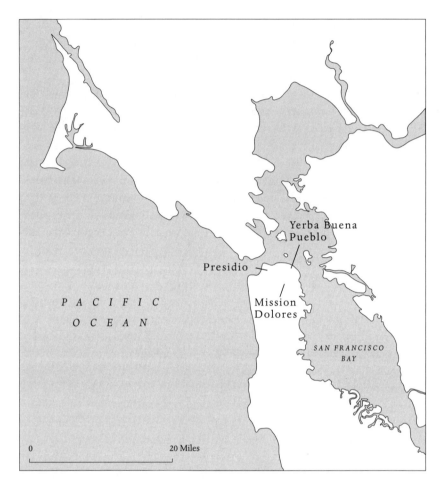

Figure 2.2
The Geographical Relationship of Presidio, Pueblo, and Mission. The topographic features at the entrance to San Francisco Bay made for an ideal location for the Presidio. Mission Dolores and Yerba Buena Pueblo were located east of the Presidio where they could be protected from attack. Both the mission and the pueblo were shielded somewhat from the winds and fog; the Presidio was not.
Source: Map by author.

Padres and Presidios in the Pacific

On September 17, 1776, the first building of the Presidio was completed, facing outward toward the vast Pacific Ocean rather than inland across the unexplored North American continent. The post was the first Spanish settlement erected in the Bay Area, predating the mission. The dedication ceremony was "accompanied by the peal of bells and repeated salvos of cannon, muskets and guns, whose roar and the sound of the bells doubtless terrified the heathen, for they did not allow themselves to be seen for many days."[23] Three weeks later, and 3 miles to the east, the mission was formally dedicated as Mission Dolores.[24]

At first, many Indians went voluntarily to the missions, keen to trade and exchange gifts. But in order to keep the missions filled with Indians, the Spanish lured them away from, and often forcibly removed them from, their villages. The Ohlones and other California Indian tribes suffered irrevocably. Because the Native American peoples had never been exposed to European diseases such as smallpox, influenza, and measles, they had no natural immunity to them. Diseases to which Europeans had developed resistance ravaged Indian populations, sweeping through the missions in epidemic waves, resulting in staggering death rates. It was not uncommon for 300 out of a population of 1,000 Indians to die during a given epidemic at Mission Dolores.[25] Imprisoned, devastated by disease, and broken in spirit, the Indians abandoned many of their traditional practices and ceremonies, thereby further disintegrating their sense of community and identity.[26] The misery in the missions induced many Indians to revolt or escape. These attempts often led the soldiers in the presidios to act to preserve order and apprehend runaway Indians.

The land occupied by the Presidio of San Francisco ceased being a home to the Ohlones. It had a new role, a new function, having been transformed from the home of a peaceful indigenous community into a fortified defensive (and occasionally offensive) space.

Bastion of Empire? The Presidio, 1776–1820

The early years of Spanish occupation at the Presidio were filled with both natural disasters and administrative friction. In the winter of 1778, the Presidio suffered heavy damage from high winds and severe storms that destroyed a major part of the walls and a warehouse.[27] A year later, fire destroyed the hospital tent and officers' quarters; this mishap was followed by another harsh winter, which caused still more damage. By 1780, few of the original buildings remained. In part, this destruction was due to the use of adobe, which was prone to disintegrate in the rain, wind, and fog. The persistent movement of sand dunes also wreaked havoc: sand routinely buried walls and pushed up against buildings, weakening support structures.

Not only natural forces but also internal friction and administrative problems challenged the new community. Many of these problems were inherent to the need to maintain a long frontier line. Food supplies were unpredictable or scarce, and food was often rationed. Shortages of firearms and gunpowder left many soldiers ill equipped for any armed conflict. The failure of soldiers' wages to keep pace with the rising cost of Old World foods and luxuries in the New World further demoralized the enlisted troops and encouraged thievery. Presidio officers had difficulty maintaining order and dealing with acts of insubordination because court-martial decisions were made by the government in northern Mexico and could take months to reach the San Francisco Presidio. The early years may be summed up thus:

> With many problems, the *grand vision* for the Presidio of San Francisco waned as the decade of the 1780s came to an end. Its defects as a barren site with harsh climate and remote location from the rest of New Spain weighed heavily against the garrison's success. In fact, the adjutant-inspector of California even advocated the abandonment of the site but his suggestion went unheeded. The need for an outpost to protect the northernmost missions and the strategic position of San Francisco Bay made it impossible to entertain the withdrawal of the troops. Yet, after more than a dozen years of precarious existence, San Francisco stood as an impotent sign of defense rather than a bastion of empire.[28]

Although the Presidio was strategically located as a guardian of the Spanish presence in Alta California, Spanish officials did not respond to repeated requests for additional resources. In 1790, a new comandante, Hermengildo Sal, reported the Presidio's neglect and poor construction. Sal's complaints and requests for more troops and money for renovations received attention only after unidentified ships began to appear outside the entrance to San Francisco Bay.[29] In 1792, the British ship *Discovery*, captained by George Vancouver, sailed into San Francisco Bay. The Presidio fired its lone cannon in salute, and Presidio soldiers rowed out into the bay to escort the English captain ashore. Captain Vancouver described what he saw when he visited the base:

> We soon arrived at the Presidio, which was not more than a mile from our landing place ... [the conditions of the Presidio] told of the inactive spirit of the people, and the unprotected state of the establishment at this port, which I should conceive ought to be a principal object of the Spanish crown, as a key barrier to their more southern and valuable settlements on the borders of the North Pacific.
>
> Instead of a city or town whose lights we had so anxiously looked for on the night of our arrival, we were conducted into a spacious, verdant plain, surrounded by hills on every side, excepting that which fronted the port. The only object which presented itself was a square area, whose sides were of 200 yards in length, enclosed by a mud wall and resembling a pound for cattle.[30]

In fact, the Presidio was so inconspicuous that Vancouver had missed it entirely, anchoring farther south along the peninsula. As a result of this visit and Vancouver's observations, the British gained valuable information about the weak state of San Francisco Bay's defenses. Ironically, it was the Englishman's observations that prompted Spain to address the condition of the Presidio. When hostility between Spain and Great Britain heightened, Comandante Sal finally received the resources he had been requesting for many years. Renovations and fortification began. Central to these renovations was the installation of additional mounted guns and cannon. Vancouver's description of the decrepit conditions at the Presidio would influence the imagination of others for the next fifty years (see Figure 2.3).

Across the continent and the Atlantic Ocean, international events helped spark Spain's interest in reinforcing its occupancy of California. As a result of a French invasion of Spain and a resumption of fighting between Spain and Great Britain, throughout the 1790s the Spanish Crown renewed efforts to fortify the Presidio and improve conditions, thus saving the outpost from collapse. However, during the same period, Indian uprisings at many of the missions in California depleted the Presidio of troops as many soldiers were moved to these missions to quell the uprisings. In 1800, the Presidio maintained a token force of thirteen foot-soldiers and five gunners. The comandante did not have enough men to operate the batteries and guard the Indian labor force at the same time, but the troops held on despite the problems.[31] The new century found all of the California presidios in poor condition. Of the San Francisco Presidio's sorry state, one historian said,

> In 1816, the San Francisco Presidio cannon was fired to welcome Don Pablo Vincente de Sola. It exploded, injuring two cannoneers. Not long afterward, the same garrison attempted to answer the seven-gun salute offered by the

Figure 2.3
The Presidio, circa 1816. This scene was drawn by Russian artist Louis Choris, who visited California in 1793. Ohlone men are being driven in the direction of a Presidio building at the harbor entrance. About three hundred Ohlone men were used as hard labor to rebuild parts of the fort after Captain Vancouver's visit in 1792. Notice how little vegetation is growing on the surrounding hills. *Source: Presidio Army Museum Photo Collection, Golden Gate National Recreation Area, National Park Service No. GOGA–1766.0001, PAM Neg. Box 7.*

> French ship *Le Héros* as it sailed into San Francisco Bay. This time two guns burst apart. In another incident, before replying to a visiting vessel's salute, the Presidio commander had to be rowed out to the ship to borrow the necessary gunpowder. . . . The presidial garrisons seemed like dress rehearsals for a comic opera.[32]

This humorous anecdote highlights the severity of the problems and inadequate support that officers at the San Francisco Presidio had to contend with.

By 1815, Indian unrest at the missions and repeated excursions by Russian fur traders, British merchants, and Americans were signaling the beginning of the end of Spanish domination. It was also the end of San Francisco Bay's relative geographic isolation. Although the Spanish government had done little to develop the San Francisco Bay Area, others began to discuss the potential of the harbor.

Spain's hold in the New World began to weaken, and by 1820 Spain's empire in Mexico had collapsed. In 1821, Mexico took advantage of Spain's increasing weakness, declared independence, and took possession of the San Francisco Presidio along with all Spanish land in North America. Revolution in Mexico ended Spain's rule in California. A new postcolonial regime in Mexico City, first an empire, then later a republic, replaced Madrid's four decades of control at the Presidio of San Francisco. In April 1822, the governor and officers of California took an oath of allegiance to the new Mexican regime.

MEXICAN OCCUPATION, 1822–1846

Despite the change in authority from the Spanish Crown to the independent Mexican government, the Presidio continued to suffer from neglect, because of the instability of the new Mexican republic.[33] All four presidios in California remained undeveloped and inadequately funded. Disarray and discontent characterized the years of Mexican occupation, for the government had concerns more pressing than buttressing outposts in its remote northern territory. In 1825, three years after Mexico assumed control over California, the Presidio troops still were using their old firearms, equipment, and uniforms from the Spanish regime.

Historical patterns of failed "grand visions" for the Presidio were repeated. In 1830, Mariano Guadalupe Vallejo, the new commander of the San Francisco Presidio, reported on the decrepit state of the buildings and requested replacement troops. The central government sent no additional resources. One English visitor commented, "Any foreign power if disposed to take possession of California could easily do so."[34] Although the Presidio had suffered visible damage, it captured the imagination as a place of promise and potential. Another visitor recalled the Presidio as "a paradise and nothing marred their [the early settlers] haven of delight until the march of civilization reached the shores of the mighty Pacific."[35]

Other maritime powers, appreciating the significance of San Francisco Bay, condemned the neglect of this potentially central harbor by its Spanish and Mexican rulers. Mexico's tenuous hold on California prompted British, French, and even Russian explorers to dream of possessing this land, which held the promise of unlimited riches. Criticism of Spanish and Mexican rule of California during the 1820s, 1830s, and 1840s was one aspect of a search for a way to justify the seizure of this outpost. The French compared California to the south of France and thought they could cultivate in California a civilized administration in the European style. The Russians envisioned extending the tsar's possessions in the Pacific and saw San Francisco Bay as a crucial port for the Russian navy. The British also coveted California, and Vancouver's enthusiastic observations revived hopes for a western British colony.[36] In addition, the presence of Americans, who had begun to voraciously carve up the western frontier, seemed inevitable. America's "Manifest Destiny" seemed to justify expansion. The spirit of Manifest Destiny, the promise of the West and its "harbor of harbors" that Father Font had described ninety years earlier, inspired the American travel writer Richard Henry Dana, who wrote of the San Francisco Bay in the early 1830s:

> If California ever becomes a prosperous country, this bay will be the center of its prosperity . . . with several good harbors, with fine forests in the north; the waters filled with fish, and the plains covered with thousands of herds of cattle; blessed with climate, than which there can be no better in the world; free from all manner of diseases, whether epidemic or endemic . . . in the hands of an enterprising people, what a country this might be![37]

By the 1830s, various interest groups had concluded that Spain and Mexico were unworthy of the land they possessed. They envisioned an alternative California. This place that had symbolized Spanish conquest now seemed pathetic and contemptible.

In 1833, the Mexican government passed a Secularization Act, taking from the missions almost all of their land and selling it to private owners, or rancheros. San Francisco's Mission Dolores was among the first to be secularized. By 1834, the Franciscan Order in California was in disarray. The dissolution of the mission system was accompanied by a deterioration of buildings, and it dispirited Indians and padres alike. Although the missions disbanded and the Indians were no longer held in formal slavery, many people continued to "sell" Indians as penal labor.[38] The Indians were freed from the missions; but a return to their traditional way of life was impossible, so they drifted to nearby pueblos or worked as ranch hands. The Ohlone descendants did not return to the Presidio area.

Winter storms and heavy rains in 1834 caused further deterioration at the Presidio. Vallejo recommended that whatever could be salvaged should be sold to provide back pay for troops.[39] By 1835, the Presidio was no longer an important garrison. Though all but deserted, it still continued operations, however minimally, as a military post.

Meanwhile, as westward expansion of the United States accelerated, interest in the San Francisco Bay Area as a home port for American whalers grew. In 1835, President Andrew Jackson first approached Mexico with an offer to buy the San Francisco Bay Area for $5 million.[40] Mexican authorities considered the offer but declined to sell the land. Despite obvious U.S. interest in Mexican territory, the Mexican government did little to change its policies toward the presidios. Mexico lacked the money or political will to operate the California territory. In 1841, Duflot de Mofras, a French visitor, commented that "the presidio of San Francisco is falling into decay, is entirely dismantled, and is inhabited only by a sub-lieutenant and five soldier rancheros with their families."[41] Later in the same year, a U.S. naval officer wrote, "after passing through the entrance, we were scarcely able to distinguish the Presidio; and had it not been for its solitary flag staff, we could not have ascertained its situation. The buildings were deserted, the walls had fallen to decay, the guns were dismounted, and everything around it lay in quiet."[42]

California in the years of Mexican rule acted "as a mirror, faithfully reflecting back all the chaos, lack of direction, and serious political instability that characterized Mexico itself."[43] The problems of the Presidio mirrored those of California—remote, underdeveloped, its status unclear in the eyes of Spanish or Mexican rulers, never having a clear purpose or priority. The earlier grand vision of the Presidio as a bastion of empire, a guardian of the riches of the Pacific, faded into obscurity. The Presidio was without troops or resources. The abandoned adobe buildings began to crumble, and by 1846 few of the structures remained habitable. It seemed only a matter of time before Mexican control of California and the San Francisco Presidio would come to an end.

CHAPTER 3

THE PRESIDIO AS U.S. ARMY POST, 1846–1906

THE GOLD RUSH

In 1846, the United States declared war on Mexico. U.S. naval forces that had been hovering along the California coast were instructed to take possession of San Francisco Bay and hoist the American flag.[1] Captain John Frémont and twelve men crossed the entrance to San Francisco Bay and occupied the undefended Presidio. The Mexican flag came down and on July 9, 1846, the Stars and Stripes first flew over the Presidio of San Francisco. During the occupation Frémont, inspired by the natural environment, named the entrance to the San Francisco Bay the "Golden Gate," a name that would persist and inspire visions of riches.[2] The Presidio began to buzz with activity. Work started immediately to repair the dilapidated buildings and to build a new road to connect the fortress and the main post. In 1847, volunteers from New York, under Major James Hardie, set to work repairing the quarters, storehouses, and many of the damaged buildings.

In 1848, the discovery of gold at Sutter's Mill changed the Presidio and its pueblo to the east forever. Shortly after the discovery of gold, Mexico formally ceded the sparsely settled California territory to the United States. The nation was infused with Gold Fever, and thousands flocked to the hills of Eldorado, including many Presidio troops who deserted their posts to search for their own fortune. Consequently, the Presidio garrison forces stood at three officers and thirteen men.[3] Despite the efforts to rebuild the post, the Presidio again lay nearly abandoned, returning yet again to historical patterns of neglect and disrepair. For the next ten years, the Presidio was as it had been during its Spanish and Mexican past: its purpose and priority were muddled; its appearance deteriorated because limited funding prevented large-scale renovations; and its work force dwindled. However, an ever-increasing trickle of Yankee migration to California had begun, and with it would come significant changes.

The Pueblo and the Presidio

The rush for gold transformed the small, scruffy, adobe pueblo of Yerba Buena, which lay on a cove 2 miles east of the Presidio, into the gateway to the gold mines.[4] Home to less than a thousand residents, Yerba Buena had been settled by Americans, Mexicans, Europeans, and a scattering of Russians. The pueblo stretched for only a few blocks in each direction, but it was well-enough established to attract commercial activity induced by the discovery of gold. Residents of Yerba Buena talked of it becoming an important city. In January 1847, a year before the Gold Rush, a local newspaper, hoping to link the growing settlement with the much better-known San Francisco Bay, had proclaimed that Yerba Buena would henceforth be known as San Francisco.[5] Historian Kevin Starr has described California during the Gold Rush years as a Homeric world of journeys, shipwrecks, labor, treasure, and killings.[6] Gold Fever and the wealth flowing into the tiny town produced a fixation on the instantaneous. The explosive growth of the city is but one dramatic illustration. Almost instantly the pueblo was transformed into a thriving city, and San Francisco became *the* commercial, financial, and shipping capital of California. Prospectors and immigrants came from elsewhere in the United States and from Europe and Mexico by the thousands. San Francisco was both a maritime boom town and a western frontier. By 1849, more than five hundred ships lay anchored off its waterfront, and the small pueblo had been transformed into a major town.[7] In 1850, more than five hundred vessels left eastern ports to join those already anchored in the bay. The white population, less than 500 in 1848, had grown to an estimated 25,000 two years later. By 1855, less than ten years after the discovery at Sutter's Mill, approximately 50,000 people lived in the new city.[8]

As a result of the flood of adventurers and the influx of Gold Rush money, San Francisco witnessed a flurry of land surveys and speculation. A rectangular grid divided the city into more than 450 parcels. In one four-day period, more than half of the 450 lots sold at auction.[9] Piers were constructed to handle the steamers, brigs, schooners, and even whalers, their occupants determined to secure their fortune in the gold mines. Groceries, saloons, hotels, and brothels arose to service the exploding population. The city quickly ran out of lots to sell. The city surveyor was instructed to survey additional land. The grid was imposed beyond the original Yerba Buena settlement, with no regard for topography. Waves of land speculation continued, pushing land prices higher and higher. Slowly but persistently, the burgeoning city of San Francisco moved southward toward the mission and westward toward the edge of the Presidio. The Presidio and the pueblo were on their way to becoming part of the same urban space.

These were critical years in the life of the infant city and the Presidio. Unsettled land titles thwarted development. Squatters were everywhere, including on Presidio territory. Claims to landownership were challenged and often resolved by violent means. The threat of encroachment on the Presidio prompted President Millard Fillmore to set aside several tracts of land, including the Presidio, as military reservations.[10] The proposed boundaries of the Presidio reservation encompassed about 10,000 acres and included the present-day post, much of the western edge of the San Francisco Peninsula, and everything northwest of a line starting at present-day Fisherman's Wharf and ending at the Cliff House (see Figure 3.1). Local citizens complained bitterly, and a month later Fillmore modified the boundaries, reducing the Presidio's acreage to 2,500. Afterward, for many years, the federal government resisted attempts by local speculators and others to acquire pieces of the Presidio reservation.[11] The threat of developing the Presidio has been a recurring theme. For more than 150 years some residents of San Francisco have looked at the Presidio and devised plans for its development and use because they envisaged it as part of the city.

The U.S. Army and the Presidio

In 1850, the U.S. government, now in possession of California, appointed a military commission to develop several military sites around San Francisco Bay to enhance the defense of the harbor. The increased military presence would serve two purposes: to maintain law and order and to protect a highly valued resource—gold. To bolster to the Presidio's defense of the Golden Gate, the commission selected Angel Island and Alcatraz for a secondary line of defense forts and lighthouses, Mare Island for a shipyard, and Benicia for an arsenal and supply depot (see Figure 3.2). The Army established Fort Point on a high promontory at the narrowest part of the entrance to the bay (see Figure 3.3). Fort Point became the lookout site for the soldiers guarding the harbor. The three additional defense posts—at Fort Point, Angel Island, and

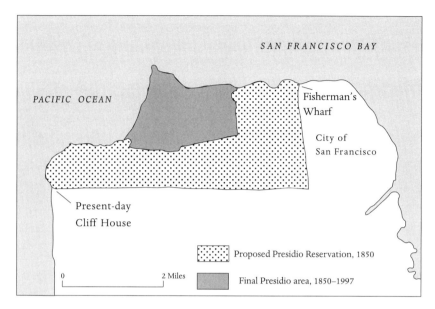

Figure 3.1
Proposed Presidio Reservation of 1850. President Millard Fillmore first proposed boundaries for the Presidio reservation to encompass about 10,000 acres. This included the present-day post, much of the western edge of the San Francisco Peninsula, and everything northwest of a line starting at present-day Fisherman's Wharf and ending at the Cliff House. When local citizens complained, Fillmore modified the boundaries, reducing the Presidio to approximately its present size.
Source: Map by author.

Alcatraz—meant that any ship entering the Golden Gate had to sail past a gauntlet of guns and cannon. The construction and fortification of these additional military posts enabled the U.S. military to take firm possession of the Bay Area. Although the numbers of troops manning the Presidio had dwindled during the Gold Rush, the post played an important role in establishing a federal presence in the new territory during the period of military government that followed the Gold Rush.

In 1854, Colonel Joseph Manfield, the War Department's inspector general, recommended remodeling the Presidio as a major priority. During the late 1850s, improvements (in addition to the construction of Fort Point) included the construction of picket fences, a medical facility, and a large multipurpose building for artillery and stores (all located at the Main Post).[12] One visitor wrote, "the old adobe buildings, and a portion of the walls are

Figure 3.2
Defending the Golden Gate. In 1850, the Army built forts and lighthouses on Angel Island and Alcatraz Island and constructed a shipyard at Mare Island and an arsenal and supply depot in nearby Benicia as part of a project to enhance the harbor's defense system. *Source: Map by author.*

there. . . . The castle of the Mexican *comandante* and the fort are now occupied by American troops; and neat, white-washed picket fences supply the place of the old walls."[13]

In the second half of the 1850s, the Presidio's troops were involved in Indian-miner conflicts as well as in the Indian campaigns in the Washington Territory.[14] Between 1859 and 1861, a number of units arrived and departed

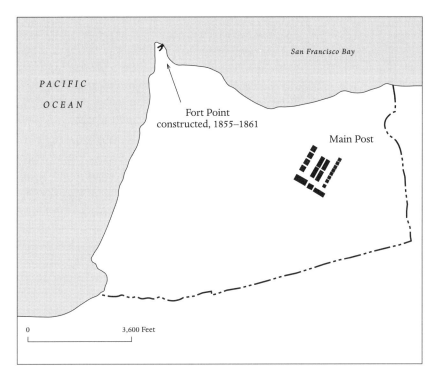

Figure 3.3
Fort Point at the Presidio in 1860. Fort Point, constructed between 1855 and 1861, was located west of the Main Post. The center of Presidio life, however, remained in the eastern portion of the post reservation. *Source: Map by author.*

from the Presidio.[15] Army engineers came to construct the other permanent coastal fortifications on Alcatraz Island. In the years after the Gold Rush, the role of the Presidio was bolstered by its unique position in relation to the city and by the necessity of establishing law and order to control the immense wealth flowing through the harbor and the gold resources in the California hills.

In a few short years, a pueblo became a mercantile city and the abandoned Presidio was renovated. In 1854, the historian Elisha Smith Capron wrote of San Francisco's metropolitan maturation: "Although the country has been thus occupied for but four or five years, it looks, in all respects as old as Massachusetts or Connecticut . . . many parts of the city have the appearance of an old town; and in passing through them, one often forgets that he is not in New York or Boston."[16] The city continued to grow by leaps and bounds, by 1860 commanding more than 2,000 acres. So it was fortuitous that Millard Fillmore had created the Presidio reservation to protect and ensure a continued military presence at the Golden Gate.

THE PRESIDIO FORTIFIED

In 1860, the Presidio stood as the central defense post for the military in California. When the Civil War began in April 1861, the Presidio's strength exploded to more than fifteen hundred soldiers, who manned the harbor defenses, marched east and south to quell secessionists, and assisted in maintaining order among the growing population in the Bay Area.[17] The Presidio army arrested southern sympathizers, administered oaths to the Union, and paraded to celebrate Union victories. Presidio troops also served under the chief of police for the city of San Francisco, maintaining law and order until the end of the war.[18] Although California's role in the Civil War was minor, symbolically these actions galvanized loyalty in the new state isolated on the western fringe of the continent.[19]

The suspected presence of Confederates in the Pacific and the fear of losing this territory to the South prompted the Army to begin new construction projects. At the beginning of the Civil War, the officers' quarters at the Presidio were wholly inadequate for an expanding military role. So in 1862, the Army constructed wood-frame cottages along the eastern side of the Presidio, facing the city (see Figure 3.4). Other new structures included enlisted-men's barracks, a carpentry shop, nine laundresses' quarters, a commissary storehouse, an artillery magazine, and four stables, which could house two hundred horses. No longer a small collection of adobe structures, the Presidio had become a substantial frontier/coastal Army post.[20] In the years after the Civil War, Union troops, chiefly arriving by sea from all over the country, were funneled through the Presidio on their way to fight in the Indian Wars of 1865–1890. However, it was the decades of peace following the Civil War that brought about the most significant changes to the Presidio.

PRESIDIO BEAUTIFUL, CITY BEAUTIFUL: GILDED AGE REFORMS AND THE URBAN ENVIRONMENTAL MOVEMENT

The post–Civil War years in America are associated with accelerated urbanization, industrialization, and the emergence of a broad set of social reforms. Americans were carving up the western frontier, the government was en-

Figure 3.4
Officers' Quarters on Funston Avenue. Built in 1862, this building is a blend of Greek Revival and Georgian Revival styles. *Source: Photograph by author.*

couraging settlement, and a few began to fear that the wilderness would soon be gone. At the same time, cities were growing; and tenement crowding, disease, air pollution from industrial manufacturing, and water pollution from new industrial development were becoming serious problems. Poor sanitation and contaminated water and air from the manufacturing processes of mills, oil refineries, tanneries, factories, and furnaces caused environmental distress and health problems among many urban residents.[21] These types of problems produced a wave of concern about environmental quality. In many cities, citizens formed groups to deal with sanitation problems, crime, disease, poverty, and pollution. Other concerned citizens, notably John Muir and George Perkins Marsh, began to publicly protest the plunder of the forests and other wild areas. Within this broader reform movement, two different environmental movements came into being: the movement for wilderness conservation (the establishment of national parks) and the urban environmental movement (reforms in urban planning and development).[22]

From 1850 to 1880, Henry David Thoreau's *Walden* and the poetry of Ralph Waldo Emerson influenced many educated elites and newly rich city residents to seek the solace and beauty of the natural world. These works, small in themselves, fostered a deepening of consciousness about the values of nature and, some have argued, created momentum for the environmental movements of the 1870s and 1880s.[23] Many of the readers of this naturalist philosophy were residents of one of the thousands of towns and cities facing increasing industrial pollution and other ills. Romantic nature writing influenced many activists in the wilderness conservation movement, but it also encouraged urban residents to redefine the natural world, not as a place of chaos and amorality, but as a place for the celebration of life and spiritual restoration. This redefinition was a profound shift in attitude. Nature was seen as the ultimate restorer and purifier of a humanity corrupted by civilization and urbanization.[24] Reformers pondered ways to bring nature to the city.

The demand for urban reform grew as citizen groups pressured for an end to the foul smells and health hazards in cities. So-called public improvement groups focused on health and aesthetics and pushed for water purification, street cleaning, and sanitation systems to improve public health; the expansion of transportation and communication networks; the establishment of housing codes; the construction of settlement houses in immigrant neighborhoods; and the extension of city parks, public squares, and playgrounds.[25] Many city planners saw developing city parks as part and parcel of enlightened and comprehensive urban planning. These reformers believed city parks and open spaces had a spiritual quality that could restore the soul of languishing urban dwellers, an economic purpose (to preserve open spaces from exploitation and rapid development), and a social purpose (to be a safety valve against malcontents who otherwise might take to the streets).[26] For the next several decades during the Gilded Age, urban reformers and the small but growing number of planners and landscape architects advocated the creation of parks as an integral part of urban development.

A Portent of Things to Come: The Presidio as Public Park?

In the post–Gold Rush years, San Francisco, like many cities, experienced the rise of a new investment class willing to invest capital in a wide variety of industrial and agricultural ventures.[27] As a result of the influence of this new class, the city entered a new cultural and social phase. San Franciscans considered themselves cosmopolitan and thought of their city as the "Paris of America."[28] But city residents—in particular this newly rich class—realized that San Francisco lacked the institutions and physical attributes of a great city. It had no large public parks, no tree-lined boulevards, no art museums, and no civic monuments. One consequence of the instantaneous creation of the city was that little time had been spent planning for public recreation, including parks and promenades. Several influential citizens embraced the ideas of the urban reformers and helped to spark a wider public discussion. Newspaper articles and editorials suggested setting aside land for recreation and leisure; it was time for a great urban park. In 1865, the *San Francisco Evening Bulletin* ran the headline "The Need for a Great Park for San Francisco" and noted that the park should be in the hills west of the city.[29] Citizen activists circulated a petition and presented it to the San Francisco Board of Supervisors in 1865. It read:

> The great cities of our country, as well as of Europe, have found it necessary at some period of their growth, to provide large parks, or pleasure grounds, for the amusement and entertainment of the people.
>
> No city in the world needs such recreation grounds more than San Francisco.... Until some provision is made to meet this need, however successful and impressive the business growth of San Francisco may be, it will not be an attractive and impressive place for families and homes.
>
> ... it would seem to be wisdom, before the suburbs of the city are more thickly populated, to have some general plan adopted for the pleasure grounds and connecting avenues, and to secure the required land as early as possible.[30]

One location suggested for this great urban park was the Presidio. Various parts of the Presidio reservation had became favorite weekend and holiday spots. Revelers in search of an antidote to urban life strolled the beaches, picnicked on the high grassy meadows, and walked along the cliffs overlooking the Pacific Ocean and harbor. Several citizens of San Francisco began a campaign to have Congress give the Presidio to the city, so it could be used for a park and for residential and business development.[31] Local newspapers and the business elite endorsed the plan. California's senators introduced bills in the U.S. Senate to transfer all but 400 acres of Presidio lands to the city; the remaining acres would be used by the War Department for defense.

The plan died, failing to muster enough senatorial support. Determined to build an urban park, the city considered other sites. Eventually, a site to the south, similar to the Presidio in its sand dunes and lack of vegetation, was chosen for Golden Gate Park. In 1870, the city sold $75,000 in bonds to begin construction.[32] In April of that year, an editorial in the *Daily Alta California* noted: "Every now and then one hears of some scheme to benefit private speculators who cast covetous eyes upon the ample government reserve on the outskirts of this city, known as the Presidio.... So far, all these designs have happily come to naught, and the reservation remains intact."[33]

Over the next hundred years, there would be other efforts to secure the use of Presidio lands as public open space.[34]

Improvements to the Presidio

With the threat of annexation and development removed, the Army initiated further improvements at the Presidio in the 1870s. Officers' cottages were built, new magazine sheds constructed, stables enlarged, and enlisted-men's barracks renovated. In 1871, the War Department recommended relocating the Military Division of the Pacific headquarters to the Presidio. The unstated purpose of the move may have been to thwart efforts in the city to acquire the Presidio lands for a park or for other nonmilitary development.[35] The Army required additional money to cover the moving and construction expenses of relocating the divisional headquarters. It took several years for Congress to approve the appropriation. Eventually, the divisional headquarters officially relocated to the Presidio, thereby bolstering the military importance of the Presidio as an administrative center. Extensive remodeling began.

Some of this remodeling was for aesthetic rather than practical reasons and reemphasized the relationship between the Presidio and the city. San Franciscans believed the Presidio was part of their urban space (a grand open park), and the Army, at least to some extent, accepted this perception, as evidenced by its efforts to create a landscape that would please residents of the city. For example, the twelve officers' cottages built in 1862 had originally been oriented so that their backyards (and outhouses) bordered the eastern gate of the Presidio property, facing the city. In the 1870s, these houses were reoriented so that their fronts faced east toward the city and their backyards and outbuildings looked out on the parade ground.[36] The reason for this dramatic change seems to have been the general's desire to present a pleasant (and orderly) view to visitors approaching the Main Post from the city.

The Army began a beautification program; troops planted pine and acacia trees along the grounds and around the residences. Non-native eucalyptus trees (which quickly grow to commanding heights) were planted on the Main Post parade ground to commemorate the centennial of the American Revolution. Other improvements included the planting of grass, lupine, and barley to stop the persistent movement of sand dunes across the Presidio as well as to give a landscaped appearance to the buildings and homes on the post. While several projects around the Presidio were aesthetic in character, others improved public access between the Presidio and the city. The Presidio Railroad completed construction of its reservation line, which ran from the officers' quarters, to the post hospital, to the intersection of Steiner and Union Streets, where it connected with a cable railroad.[37] These improvements were signs of the Presidio's increased stature as a military base. In 1881, one general remarked: "It is scarcely necessary to say that the most important point on the Pacific Coast is the harbor of San Francisco, and that in this harbor by far the most important and valuable military possession and position is the Presidio of San Francisco."[38]

The architectural styles of buildings at the Presidio became increasingly diversified as the Army adopted new standard designs. Throughout the 1880s, materials used in buildings included brick, concrete, and hollow tile, in addition to wood. Like other federal buildings constructed at the time (such as post offices, courthouses, and state capitals), military installation buildings were in either Colonial Revival or Classical Revival style.[39]

While improvements were being carried out at the Presidio, San Francisco planners initiated several civic improvements. Marshland around the city was drained and filled for construction. The city continued its westward and southward expansion, aided by new civic projects that improved transportation (railroads, horsecar lines, and cable cars), thereby making formidably steep hills accessible. Transportation improvements spurred the development of new residential districts. Pacific Heights was built, bordering the eastern boundary of the Presidio. This wealthy neighborhood had the open space of the Presidio as its backyard. By the 1880s, San Francisco had begun work on Golden Gate Park, developed a thriving hotel and restaurant industry, and constructed the third largest opera house in the country.

Presidio Beautiful

Through the turn of the century, significant changes occurred in the city of San Francisco and the Presidio. These changes beautified both places, and the planning decisions that led to them had a profound impact on the cultural landscape.

Throughout the country during the 1880s, urban reformers called for cities to create and develop parks and open spaces. Landscape architects were employed to interpret nature's beauty and to design great parks for cities. Frederick Law Olmsted, the most influential landscape architect, participated directly or indirectly in nearly every major urban park project and established new standards of landscape design.[40] The focus on the creation or improvement of urban parks reflected a new environmental ideology.

Park projects also represented an economic tribute to nature worship and romantic-era landscape ideals. Between 1858 and 1873, New York spent nearly $14 million to construct Central Park; and between 1870 and 1900, Boston spent more than $25 million on its city park system.[41] San Francisco spent nearly $350,000 in the first six years of construction on Golden Gate Park.[42] However, not only official city-owned parks were being affected by the new ideas. Olmsted's intellectual legacy to the urban park movement was his conviction that beauty and utility were inseparable. This idea was behind plans to improve the Presidio.

The Presidio is exposed to the strong winds blowing in from the Pacific. The winds continually swept fog and sand across the site. Sand dunes, a perennial problem, could be fiercely destructive, burying buildings and walls. The need to halt drifting sand provided one impetus for landscape improvements at the Presidio. The planting of trees, grass, and shrubs would create effective windbreaks, anchor sand dunes, and stem erosion. Tree planting also addressed aesthetic concerns. It would create a landscape on which the citizens of San Francisco could gaze and find beauty. Trees would mitigate the Presidio's reputation as a cold, windswept place.

Beautification of the Presidio mirrored the urban park movement, which embraced civic improvement with regard to aesthetics, parkways, tree-lined boulevards, public squares, and above all, deliberate and comprehensive park planning. All of these improvements were part of the Presidio beautification and forestation plan that took shape in 1883, conceived by Major William Albert Jones, an Army engineer. Of his plan to beautify the Presidio, Jones wrote:

> The main idea is to crown the ridges, border the boundary fences, and cover the areas of sand and marsh water with a forest that will generally seem continuous, and thus appear immensely larger than it really is. By leaving the valleys uncovered or with a scattering fringe of trees along the streams, the contrast of height will be strengthened. *In order to make the contrast from the city seem as great as possible, and indirectly accentuate the ideas of the power of the Government*, I have surrounded all the entrances with dense masses of woods [my emphasis].[43]

Jones planned to make the Presidio seem bigger and more forested than it actually was. He wanted to imprint the place with the authority of the federal government and remind everyone of the military ownership of the land.[44] San Franciscans supported the plan, as did local members of Congress.

Although urban park planning was nothing new by 1883 (Golden Gate Park and New York's Central Park had been constructed earlier), the Presidio forestation plan was significant because it was the first large-scale (and public) landscaping effort of its type by the Army. Prior to 1880, military improvements and construction projects consisted of defensive fortifications and various support structures, such as barracks, warehouses, and command centers. The beautification of the Presidio had no direct military purpose; it did serve authority, however. The project would change the physical landscape in an enduring way.

The "Eucalyptus Rush" and the Presidio

The tree-planting program began in 1886, when San Franciscans celebrated the city's first Arbor Day at the Presidio. Local philanthropist Adolph Sutro donated 3,000 trees, and more than four thousand women and children, along with Presidio troops, planted them around the perimeters of the Presidio. In 1893, 60,000 indigenous Monterey pines, a shoreline tree, were planted in an area of 40 acres along the western borders of the Presidio, to provide a windbreak. The year after, with the Army's permission, the San Francisco Golf Club built a nine-hole course within the southern boundary of the Presidio. Presidio troops helped plant pine and cypress trees along course fairways and boundaries as windbreaks. In all, the Army planted nearly 100,000 trees on the Presidio between 1888 and 1897 at a cost of $58,000 (see Figure 3.5).[45] The species included live oak, Monterey pine, Monterey cypress, maritime pine, shore pine, Australian blackwood, English oak, English elm, ginkgo, magnolia, Kentucky coffeetree, giant redwoods, pepper trees, and lawson cypress.

Non-native trees and plants changed the Presidio's natural environment. Australian eucalyptus became the dominant arboreal presence on the post. In California, landscape architects had become enchanted by the eucalyptus because of the distinctiveness of the light filtered through its leaves.[46] Natural historian and San Francisco resident Harold Gilliam has termed the 1880s the "Eucalyptus Rush." A statewide eucalyptus-planting program promised to end the U.S. hardwood famine. At the time promoters of the eucalyptus believed it would become the principal source of hardwood and produce greater wealth than the burgeoning orange business.[47] Around California, the Presidio included, eucalyptus boomers planted the "magic new money tree," and soon the once-barren and scrub brush hills were clothed in rapidly growing forests.[48]

Figure 3.5
Presidio Tree Planting, circa 1887. This photograph of the Presidio was taken from the city, facing west. Notice the rows of young trees planted in the early 1880s along the eastern perimeter of the post. They are about ten years old. Trees were also planted at the officers' quarters on Funston Avenue (the road in the foreground of the eastern part of the picture). The tree-planting program was part of the beautification of the Presidio. *Source: Presidio Army Museum Photo Collection, Golden Gate National Recreation Area, National Park Service No. GOGA–1766.0001, PAM Neg. Box 1, Presidio Army Museum Photo Print Collection, Box 1, Folder 4.*

In addition to the forestation program, the Army made plans to fill in the 110-acre swamp and marsh along the bay-front side of the Presidio, in an effort to reclaim more land. These plans clearly were efforts to create a distinctive Presidio landscape.

Efforts to make the Presidio a distinct and separate place without closing it off from the city and its residents are perhaps best illustrated by the Army's construction of a 4-foot-high stone wall along the Presidio's southern and eastern boundaries in 1895 (see Figure 3.6). In addition, in 1896, the Army constructed several gates at key Presidio/city access points.

The trees and walls affirmed the power of the property owner and political power.[49] The wall was a statement of territoriality and power. It marked a federal, military presence in a dense urban area. Perhaps it also conveyed a second message. Although the wall demarcated federal ownership of land, it did not create an absolute demarcation, as the walls and fences around many military bases do. The Army constructed the wall to keep squatters from moving onto the open land, but the wall also served to preserve areas of the Presidio as open space—open space that remained available not just to the military but to all city residents to use. Since the wall is only 4 feet high and contains several gates for public access, it neither dominates the landscape nor keeps people out of the post. Rather, its low height and gates are visual symbols that demarcate a fluid boundary—from urban to park. The wall thus defines the Presidio as a landscape that is distinct from

Figure 3.6
The Presidio Wall. Built between 1895 and 1896, the wall stretches along the eastern and southern boundaries of the post. Only 4 feet high, the wall has not been a true barrier. The Presidio is to the left of the wall. *Source: Photograph by author.*

the city. Generally, the Presidio gates were open to the public and not guarded (see Figure 3.7).

What is significant about the beautification of the Presidio is that much of it was for public consumption. The beautification project created a parklike landscape and, despite the walls and gates, reinforced the identity of the Presidio as public open space. Plans for Presidio beautification shared many of the same ideas and inspirations as the building of Golden Gate Park. In fact, most of the plants used in the Presidio beautification project had been used in Golden Gate Park: grasses, lupines, Monterey cypress and pines, eucalyptus trees, and acacias. Presidio beautification and the construction of Golden Gate Park were similar to many efforts in other cities around the country to preserve and develop large natural areas within urban boundaries.[50]

City Beautiful

The explosive growth of the city of San Francisco during the Gold Rush and in the decades following had been unplanned. The standard rectangular grid had been imposed on a dramatic topography with little discussion, and dur-

Figure 3.7
The Presidio Gate, circa 1906. This photograph shows the entrance from Presidio Boulevard (the South Gate). Notice how tall the cypress trees, planted in the 1880s, have grown along the southern Presidio perimeter. Guards were posted at the gates for parades and ceremonies and during World War II, when fear of Japanese submarine attacks closed the Presidio for the first time in its history. *Source: Presidio Army Museum Photo Collection, Golden Gate National Recreation Area, National Park Service No. GOGA–1766.0001, PAM Neg. Box 1, Presidio Army Museum Original Photo Collection, Box 1, Folder I.F. Also courtesy of San Francisco History Center, San Francisco Public Library.*

ing the early years of growth, San Franciscans had displayed a talent for urban mismanagement.[51] The city sold the bulk of its public land to speculators, who divided it into dreary rectangular lots, displaying a total lack of geographic sensitivity.[52] By the turn of the century, San Francisco was becoming one of the most densely populated cities in the United States. Some believed it was a crucial time: San Francisco would be either great and beautiful or great and ugly. So many people began to discuss ways to improve urban life. This turn-of-the-century urban reform movement—the City Beautiful movement—was led primarily by middle-class and upper-class urban Americans who wanted to synthesize cultural, aesthetic, political, and environmental ideas in order to refashion cities.[53] City Beautiful advocates renewed the spirit of civic life, continuing many of the reforms begun in the decades after the Civil War. Reforms involved projects such as municipal art (sculpture and street art), boulevards, parks and playgrounds, lighting, sewers, street paving, sanitation and transportation improvements. Many City Beautiful advocates were environmentalists who hoped to harness the influence of "nature's beauty" to shape social and urban reform.[54] Around the country, in cities such as Washington, D.C.; Chicago; Dallas; San Diego; Raleigh, North Carolina; and Harrisburg, Pennsylvania, civic organizers planned urban improvements.

By 1900, in cities across the United States a myriad of civic organizations had developed to tame the disorganized, wildly growing cities and to restore a sense of community. Many of these civic groups were "improvement associations," which worked to promote a wide range of civic improvements. By 1904, nearly twenty-five hundred civic improvement societies were supporting the American planning movement.[55] In that same year, San Franciscans formed the Association for Improvement and Adornment of San Francisco. Their goals were

> To promote in every practical way the beautifying of the streets, public buildings, parks, squares, and places of San Francisco; to bring to the attention of the officials and the people of the city the best methods for instituting artistic municipal betterments; to stimulate the sentiment of civic pride in the improvement and care of private property; to suggest quasi-public enterprises; and, in short, to make San Francisco a more agreeable city in which to live.[56]

Many San Franciscans envisioned the city as more than just the economic capital of California. They wanted it to be unsurpassed culturally, socially, and environmentally. Association members endorsed the principles of the City Beautiful movement. In one of its first official actions, the San Francisco Improvement and Adornment Association (consisting primarily of members of the urban elite) asked Chicago architect Daniel Hudson Burnham to design a plan for San Francisco similar to the one he and Frederick Law Olmsted, Jr., had proposed for Washington three years earlier.

For many years Olmsted's firm had employed America's most prominent landscape architects, and Burnham had become the most influential urban planner of the City Beautiful movement. Together, Olmsted and Burnham guided the formation of many City Beautiful ideals. Because of Olmsted's intellectual legacy, there are several points of continuity between the urban park movement of the 1870s to the 1890s and the City Beautiful movement of the early twentieth century, particularly with regard to the role of parks in urban spaces. These points of continuity included the view of urban parks as the "lungs" of the city and its "breathing places"—in short, the antithesis of urban artificiality and its attendant congestion.[57]

City Beautiful differed slightly from the earlier urban reforms that had influenced beautification of the Presidio, because City Beautiful proponents thought city planning needed to be more comprehensive. They encouraged planners to understand and synthesize the various parts of the city—from parks to financial districts to residential areas—in order to blend beauty and utility. In 1904, San Franciscans began again to consider the link among existing open spaces (of which the Presidio was one) and other urban sites.

Burnham proposed numerous improvements within a geographically comprehensive plan. Like many other City Beautiful planners, he emphasized classical, monumental architecture and harmonious, symmetrical relationships among buildings. Burnham envisioned San Francisco as an "imperial city," with neo-baroque architecture that would convey grandeur and authority and with radiating boulevards and great parkways leading to open green areas, including the Presidio.[58] Parks were high on Burnham's list of recommendations, for he felt the city lacked sufficient recreation space. Burnham noted this in his plan: in the city as a whole there were approximately 286 persons for each acre of park; the average for "the most important cities of the United States," however, was 206 persons for each park acre, and Boston had set a standard of 42 persons per park acre.[59] Burnham was inspired by the hilly terrain of the city and felt it was well suited for parks and open spaces. Indeed Burnham's plan for San Francisco was primarily a plan for streets and parks; he envisioned a San Francisco in which an entire third of the city would be in parks.[60] City Beautiful advocates endorsed Burnham's ideas, convinced that his plan bespoke the value of urban design and displayed the enlightened attributes of humanizing the landscape.

City Beautiful and the Presidio

While City Beautiful ideals were influencing planners in San Francisco, they were resonating with Army engineers at the Presidio as well. By the turn of the century, the western frontier had been settled, and the U.S. military purpose was no longer to secure domestic boundaries and protect new settlements. A new mission was emerging overseas. In 1898, the United States declared war on Spain over Cuba. Thousands of troops gathered at the Presidio prior to shipping out for the Spanish-American war and the Filipino insurrection. Once again, war and conflict bolstered the Presidio's military importance. Troops arrived at and departed from the Presidio. Increased activity necessitated the construction of more barracks and the enlargement of existing hospitals, commissaries, and kitchens and messes. In addition, the Presidio strengthened its coastal defense system against possible attacks from Spain by installing a series of large concrete gun batteries on the bluffs overlooking the bay and the Pacific.

The new century dawned on a Presidio that had grown in troop strength (from 35 officers and 506 enlisted men in 1890 to 42 officers and 1,330 enlisted men in 1905). It also had grown in size and number of buildings. And it had developed a distinctive cultural landscape: intentional beautification efforts had changed the natural environment and brought into being an intensely humanized landscape.[61] The Presidio had become both a park and a small military city within the larger and continually growing urban metropolis of San Francisco.

By 1900, developments at the Presidio were paralleling those of the city. The continuity of City Beautiful ideals and attention to natural landscapes inspired continued improvements at the Presidio. At the time civic elites were asking Burnham to develop a city plan, Army engineers at the Presidio were consulting with Gifford Pinchot, head of the Forest Bureau and influential leader of the conservation movement, on landscape engineering. They also contacted various landscape architects and planners, including Daniel Burnham, to develop a comprehensive Presidio plan. Burnham recommended that the drives and concourses around the Presidio be arranged so that the public could enjoy the best views of the landscape.[62] Burnham's suggestions for the Presidio were in line with the improvements he suggested for the city; they were designed "in connection with the plans for the beautification of San Francisco." The Presidio would be *in* the city but not *of* the city.

Daniel Burnham's plans for a new majestic San Francisco were delivered to City Hall on April 17, 1906, so that they could be studied and available for public review. City Beautiful advocates were nearly giddy in their optimism about implementing Burnham's plan. The next day, everything changed.

CHAPTER 4

THE PRESIDIO IN THE TWENTIETH CENTURY, 1906–1989

THE GREAT 'QUAKE

At 5:12 A.M. on April 18, 1906, the San Andreas fault slipped. The ground rattled, rolled, and rumbled, and the destruction of San Francisco began. The earthquake itself accounted for only about 20 percent of the ruin of San Francisco; fires from hundreds of broken gas mains and gas lanterns soon set the city ablaze in a superheated firestorm.[1] For three days and two nights, the fire raged, burning more than 500 square blocks and destroying more than 28,000 buildings. The city lay in ruins. San Francisco mayor Eugene Schmitz enlisted the aid of the Presidio's commander, General Frederick Funston, to patrol the city, prevent looting, and provide relief operations (see Figure 4.1). On the post's open spaces tent cities were erected to provide temporary housing for an estimated 16,000 to 70,000 refugees.

The Presidio suffered comparatively little damage because it rests on bedrock. Damage to buildings consisted mostly of collapsed chimneys and roof damage. However, two of the three original Spanish adobe buildings were in danger of falling and too badly damaged to repair. Both were condemned and torn down, leaving only one adobe building from the Spanish/Mexican era, the Officers' Club, at the Presidio.[2]

For a handful of civic leaders and intellectuals, Burnham's plan became the symbol of recovery. An even more beautiful city would rise from the ashes—or so City Beautiful advocates hoped. They began lobbying for a rejection of the rectangular grid plan and for a revival of the promise of planning that had all but been ignored since Gold Rush days. Other residents,

Figure 4.1
Presidio Troops Aid the City, 1906. These weary troops posed for a photographer at Portsmouth Square against a background of destruction caused by the great earthquake and fire. *Source: Golden Gate National Recreation Area Park Archives and Records, Photograph No. 54.*

however, protested against anything that would delay reconstruction. They argued that much of Burnham's plan was not practical. They also worried the plan would be too expensive to implement and believed it would burden property owners with additional taxation. "Let us have a city beautiful but within our ability to pay," urged an editorial in the *San Francisco Bulletin*.[3] Opponents of Burnham's plan argued that utility was more important than beauty in solving the problems of the ruined city. After some public debate, city officials decided to rebuild San Francisco as quickly as possible, following the old patterns. Beautification and grandeur would have to wait for another generation; Burnham's ideal gave way to instantaneous restoration.[4]

Architecture at the Presidio from 1906 to 1920: A Return to the Spanish Heritage

Local architects favored the traditional and distinctly European architectural variations of Queen Anne, Edwardian, and Colonial Revival styles. Elsewhere in California, however, people began to prefer a more vernacular and Latin-historical style. In many communities in California the Spanish–Colonial Revival style became popular. It fostered intimacy and the outdoor-indoor lifestyle and harmony with the landscape so dear to Californians, and its unpainted exteriors, exposed beams, textured walls, patios, and terraces brought to mind the old Spanish missions and presidios.[5] Undoubtedly the Spanish Revival romanticized Old California and its missions and presidios. Nevertheless, the style was historically appropriate and, more important, according to its advocates, created a visually romantic environment. The desire to project a romantic image also appears to have played a role in the Army's decision to introduce the Mission Revival style to the Presidio.[6] Many of the new buildings and officers' quarters built during the next twenty years were designed in the Spanish Revival style. Perhaps the most pronounced effort to associate the Presidio with its Spanish/Mexican heritage was the design of a second fort, Fort Winfield Scott.

By 1907, the Presidio's troop strength had grown from 1,330 enlisted men to approximately 2,000. General Funston wrote to the War Department, describing the inadequacies of the enlisted-men's housing and the need for additional buildings. In 1908, construction began on Fort Winfield Scott (plans for the fort had been completed several years earlier). Fort Scott would become the focal point of the western half of the Presidio; the Main Post remained the focus of the eastern half (see Figure 4.2).

Construction had been delayed while the Army planners debated archi-

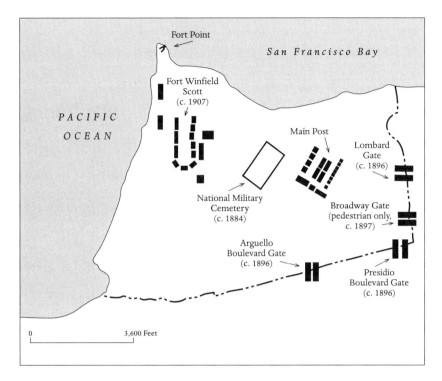

Figure 4.2
The Presidio Grows. Fort Winfield Scott, constructed in 1907, was located in the western portion of the post reservation. Several gates allowed public access. *Source: Map by author.*

tectural style. Original plans called for the Spanish Mission style, a dramatic difference from the style of Queen Anne and Colonial Revival dominating the Main Post. Some Army officials disagreed with the introduction of vernacular architecture and thought the buildings should follow standard federal styles, which emphasized universality and central authority. But critics of these European styles decried the buildings as "old fashioned, too near the ground and shabby."[7] After several years of debate as to which architectural style should prevail, Army inspector general Major C. A. Devol wrote that "the plan of buildings in the old Spanish style with tile roofs appears to be a good one, and the plan should be an ornament to the Pacific Coast."[8] Cream-colored stucco (an updated version of adobe) and red-tiled roofs with curving parapets highlighted Fort Scott's Spanish style. Fort Scott was designed in a unified Mission Revival architectural style and remains the most coherent set

Figure 4.3
Mission Revival Style at Fort Winfield Scott. *Source: Photograph by author.*

of buildings and space on the Presidio (see Figure 4.3). The construction of Fort Scott marked the first effort to "Hispanicize" the Presidio; this would be an architectural theme that would be repeated in later developments on the Presidio.[9]

Until the construction of Fort Scott, traditional construction on domestic Army posts, including the Presidio, followed standard plans rather than the influences of local architectural styles. The decision to build in the Spanish Revival style at the Presidio is noteworthy, for in all of Army history this was one of the few times and places in which construction departed from standard Army post planning. The decision to design in the Spanish Revival style was, in the case of the Presidio, an acknowledgment of the earlier post architecture of the Spanish era. The decision was far-reaching, for it accentuated the historic character of the Presidio and factored greatly in its future designation as a national historic landmark district.

Other construction between 1907 and 1915 included a bakery, five new stables, more noncommissioned-officers' quarters (in Georgian Revival style), an octagonal bandstand, an electrical substation, a coastal artillery post, and an officers' recreation center.

The Presidio and the City: Shared Space and New Projects

In 1911, the Army signed a ten-year lease with the city of San Francisco, allowing the city to reclaim the marshland in the northeast corner of the Presidio in order to construct a site for the Panama-Pacific International Exposition. From 1911 to 1914, more than 114 acres of the lower western Presidio were filled in preparation for the Expo, a world's fair to honor the completion of the Panama Canal and to symbolize San Francisco's rebirth following the devastating earthquake of 1906. Indeed, the Expo became a coming-of-age party for both the city of San Francisco and the state of California.[10] Of this self-congratulatory fervor, historian Tom Cole says, "even before the fire, San Francisco had been in the mood for a giant celebration of its own good luck at being itself . . . it has never been a city to wallow in false modesty."[11] San Francisco was a city rebuilt, now matured beyond its frontier past, and California itself was an expression of Progressivism. The Panama-Pacific Exposition was yet another example of the intertwining of pueblo and presidio. During the fair, troops from the Presidio participated in parades and ceremonies on the fair grounds and generally "lent its impressive appearance throughout the year."[12]

The First World War and the Automobile Boom: Bolstering the Presidio

In the spring of 1917, the United States declared war on Germany, and the Presidio underwent dramatic alterations. The start of the war forced the Panama-Pacific Exposition to close early. After the close of the exposition, portions of the reclaimed land reverted to the military and in just a few months became the site of Crissy Field, the air base established during the First World War (see Figure 4.4). Large camps of temporary buildings were also constructed where the Panama-Pacific Exposition buildings had stood only a few months before. Various regiments received basic training at the Presidio, and the post served as the departure point for troops being sent overseas. More buildings were constructed to handle the vast increase in troops and operations. Desperately needed warehouses were built, as was a balloon hangar when Fort Scott acquired a balloon company.

The "war to end all wars" stimulated both the city and the Presidio. It brought prosperity, federal investment, and a heightened sense of importance to both places. Later, in 1920, the War Department authorized $1 million to

Figure 4.4
Crissy Airfield, circa 1933. Air defense came to the Presidio in 1921. The planes of Crissy Field served the military in eight western states. The city is to the southeast of the photograph. The semicircular building in the upper right is the present-day Exploratorium built for the Panama-Pacific Exposition of 1915. *Source: U.S. Air Force, National Air and Space Museum via the Presidio Army Museum Photo Collection, Golden Gate National Recreation Area, National Park Service No. GOGA–1766.0038, P.J. Lloyd 30th Infantry Album.*

enlarge the runway at Crissy Field to accommodate the newer airplanes, which needed more space for takeoffs and landings. The runway was lengthened, hangars and other support facilities were constructed on the field, and pilots' quarters were built on the hill above the field. The new airfield enhanced the importance of the Presidio as Pacific Defense headquarters.

At the end of the war, the Presidio approximated its present form and size (see again Figure 1.1). By 1925, however, the Presidio's size was problematic—there was no space in which to expand to allow the new aviation technology to operate. Bounded by the city on the east and the steep cliffs of the Pacific on the west, Crissy Field quickly became obsolete despite considerable investment in improvements just a few years earlier.

During the 1920s, motor vehicles captured the American imagination. The volume of new autos demanded new roads, and the emerging automobile culture had a significant impact on patterns of urban development and land use.[13] The automobile boom changed both the Presidio and the city. The increased use of automobiles, especially for pleasure trips, profoundly affected the physical environment as miles of new roads and highways were constructed. Automobile use stimulated new commercial enterprises, created new employment, and changed patterns of recreation and exploration. Many families opted for a Sunday-afternoon automobile trip rather than the park promenade. The automobile allowed access to new scenic areas around the city and the surrounding foothills. As an echo of City Beautiful ideals, planners designed residence parks—wide-street, tree-lined residences—in the sparsely settled southwestern districts of the city. These new residential developments were to take advantage of the hills, terraces, and open spaces. Engineers completed a railway tunnel through Twin Peaks, a hilly geographical barrier west of the city center, allowing new areas to be developed. Also during this time, a citywide boulevard system took shape. State engineers constructed the road called El Camino Del Mar to link the Presidio with the Great Highway and the esplanade along Ocean Beach. The federal government constructed several other scenic drives through the Presidio as well. As a result of these various projects, several drives were linked, connecting both residents and tourists to urban parks and open spaces, starting at the Presidio to Lobos Creek, then going through Lincoln Park to the Cliff House, and down the Great Highway along the Pacific Ocean.

The formation of a metropolitan regional park system was discussed in local newspapers and among local citizen groups. As people developed a new interest in the natural beauties of the Bay Area, they began to realize that unless they took steps to preserve outstanding wilderness areas, these might be destroyed by careless private owners or misguided commercial exploitation.[14] At first there was much interest in creating a park system, but it fizzled with the stock market crash of 1929.

The Great Depression and Presidio Development Projects

On Friday, October 28, 1929, the New York Stock Exchange collapsed and a world economic crisis began. The Great Depression engulfed the nation for the next twelve years. The city of San Francisco, like cities elsewhere, suffered. On the waterfront, shipping all but disappeared; Montgomery Street (the Wall Street of the West) was silent; thousands lost jobs; and periodic labor strife exacerbated economic distress. Despite economic woes, the city possessed remarkable energy, especially for new development projects that promised to stimulate the local economy.

Proposals to close or change the use of the Presidio were raised several times in the 1920s and 1930s.[15] One proposal suggested converting the entire Presidio into an Army hospital complex. In 1927, local interests launched a campaign to sell the Presidio to real estate agents who were keen to subdivide the prime lands.[16] The Presidio seemed to be on its way to being declared surplus property. But it was saved by Roosevelt's New Deal program and federal investment in various construction projects.

Massive unemployment, associated with the severe economic distress, had two impacts on the Presidio. First, the number of military enlistees rose dramatically as unemployed young men sought refuge from dismal employment opportunities by joining the Army. As enlistments rose around the nation, the Presidio gained increased responsibilities as a Pacific headquarters post. In 1930, Congress authorized the construction of quarters for forty noncommissioned officers at the Presidio; these were built between 1932 and 1939. Second, several public relief projects took place on the post toward the end of the decade. As part of President Roosevelt's New Deal program to provide economic relief, the Works Progress Administration (WPA), Civilian Conservation Corps (CCC), and Civil Works Administration (CWA) brought additional "armies" of public workers to the Presidio to coordinate several construction projects and park improvements.[17]

At the Presidio, the focus of federally sponsored projects was the construction of the Golden Gate Bridge, but WPA workers did extensive remodeling of several buildings, including the Officers' Club, the post chapel, and the post theater. During the 1930s, more than $100,000 was spent on Presidio parkways and streets, and WPA funds also paid for improvements in utilities and telephone communications.[18] The Civilian Conservation Corps constructed new hiking trails and picnic areas and supported the planting of more than 15,000 trees on the Presidio golf course.

The WPA remodeling of the Officers' Club, the site of the only original Presidio building, was in the Spanish Revival style, and efforts were made to recover some of the original adobe portions of the structure. It reopened in August 1934, and one officer extolled "the old building . . . [which] sheltered since 1846 officers of all the branches . . . [was] the social center of the post at the cross-roads to the Orient, Hawaii, the Philippines, Cuba, Panama and Alaska."[19] The Officers' Club was the oldest building on the post, and the Army declared it a landmark because of its link to the Spanish/Mexican past. The Depression-era construction, from the 1930s to the early 1940s, marked the last major use of Spanish-derived architectural styles on the post.

Figure 4.5
Construction of the Golden Gate Bridge, circa 1935. In the foreground are the officers' quarters and pilots' houses. Notice again the absence of vegetation. Eventually, the bridge lookout spot, located on the bluff above Fort Point (under the anchor of the twin pillars) became one of the city's most visited tourist destinations.
Source: U.S. Air Force, National Air and Space Museum, and the National Archives via the Golden Gate National Recreation Area, National Park Service No. GOGA—2224, Crissy Field Study Collection, Box 2, File No. 1.

The Presidio also was saved by the construction of the Golden Gate Bridge, which began in 1933 (see Figure 4.5). The bridge would be the largest and highest single-span suspension bridge of its time and would span what has been called one of the most romantic approaches anywhere.[20] Plans had been under way for many years; however, opposition had delayed construction. Various interest groups (such as ferry-boat operators and Presidio neighbors) were concerned about the effects of such a bridge on scenery and military defense, as well as how earthquakes would affect it.[21] The construction of the Golden Gate Bridge aided the city during a period of economic depression; indeed construction of both the Bay Bridge and the Golden Gate Bridge dominated the city's imagination and energy in the "Dirty Thirties."

The construction of the Golden Gate Bridge substantially changed the Presidio's appearance and geography.[22] The approaches to the bridge were to lie on Presidio land. Army commanders voiced opposition to the project because they perceived possibly negative effects on military defense, but in the end the secretary of war overruled these objections. In exchange for the use of Presidio land and as compensation for the several structures that had to be relocated or demolished to make way for the bridge, the Bridge District constructed a number of new buildings on post. Total costs for these improvements at the Presidio amounted to $575,000.[23] The approaches were Doyle Drive, an elevated approach, which runs along the far southern edge of Crissy Field; and Park-Presidio Boulevard, which runs north to south, linking up with the southern areas of the city, notably the Richmond and Sunset districts. The building of the Golden Gate Bridge further solidified the public character of the Presidio post.

When completed, the bridge was heralded as an example of poetic engineering. Writers inspired by the Golden Gate Bridge called it "a steel harp," "the bridge that sings," "the span of gold," "an extraordinary rarity," and "a work of man that enhances rather than detracts from the work of nature."[24] The southern anchor point of the bridge became one of the busiest tourist areas in the city, and visitors took advantage of their excursion to the bridge to picnic in the Presidio or to hike or walk through surrounding Presidio trails.

THE SECOND WORLD WAR

When the Japanese attacked Pearl Harbor on December 7, 1941, the Presidio was designated the Western Defense Command center, and it functioned as a staging area for American forces throughout the war. During the war, over 1½ million military personnel and 23 million tons of war supplies passed through the Golden Gate. In the city, the number of factories engaged in war-related work grew by 33 percent, and the work force doubled.[25] New shipyards became frenetic centers of activity, building cruisers, destroyers, and aircraft carriers. By the summer of 1942, the San Francisco Bay Area was a "major arsenal of democracy."[26] It had, almost quietly, become a large complex of military establishments, of which the Presidio was administrative headquarters. The Army commanded the Presidio, Fort Mason, Fort McDowell on Angel Island, a supply depot in Oakland, and Fort Barry and Fort Baker in Marin County. The Navy operated a new base at Treasure Island, shipyards at Hunter's Point and Mare Island, and a supply depot in Oakland. Between 1941 and 1944, the Bay Area received more than $4 billion in war supply contracts and was first in contracts for ships.[27]

From the days of the Spanish occupation, when a footpath led from the

Main Post to Mission Dolores, the Presidio had been considered an "open post" by residents of San Francisco. The Army had maintained this policy in early years, allowing civilian neighbors to graze cattle on what was initially grassland and later allowing the public to use unoccupied portions of the base for recreation.[28] But fears that Japanese submarines would slip through the Golden Gate and attack the city prompted the Army to close the Presidio to civilians. Historically a public space, it was for the first time placed off-limits. Sentries guarded each of the Presidio gates; barbed wire, machine gun emplacements, and anti-aircraft guns were located on the golf course, Baker Beach, and elsewhere.

As during previous military conflicts, the Presidio assumed a heightened importance in the military's Pacific defense planning. Construction and conversion efforts were renewed. Fort Winfield Scott, which had languished in the years after World War I, became the headquarters of the coastal defenses of San Francisco. A hangar at Crissy Field was converted into a foreign language training school for second-generation Japanese-American soldiers who served as translators and interrogators for the Pacific theater of operations. Letterman General Hospital became the nation's largest debarkation hospital, handling thousands of wounded and sick returning from the Pacific.[29] The Presidio's harbor defense system, which had lain dormant since the First World War, was reactivated. But despite these important support services, the Presidio played only a minor military role in the Second World War. It did not even serve as a training ground as it had during the First World War. It became clear that the post was no longer critical to the U.S. military, although it was still considered an important administrative headquarters.

In the years after the war, the most defining moment for the Presidio would occur.

Postwar Presidio

In 1945, World War II still raged on, but Allied victory seemed imminent. Forty-six nations met at a conference in San Francisco to draft the charter of the United Nations Organization in April, May, and June of 1945. President Harry S Truman proposed the Presidio of San Francisco as the site for the United Nations headquarters.[30] The Soviet Union objected strongly to the headquarters being located anywhere on the West Coast, and when Nelson Rockefeller offered to donate land in New York City, the Presidio was bypassed.[31] Real estate speculators and others who recognized that the Presidio was becoming increasingly obsolete, not only as a defensive post but also as an Army administrative headquarters, then campaigned for the Army to abandon it so that the area could be developed.[32] Many of these local real estate speculators hoped to erect high-rise apartments.[33] During 1947, local newspapers ran several articles describing plans to develop the Presidio. The *Chronicle* detailed one plan to fill 320 acres in front of the lower Presidio in order to construct twelve thousand apartments; the mayor supported the plan, but the Army opposed it. There was an ongoing debate over whether the Presidio should be developed for housing or preserved as a national monument. By the 1950s, both leading papers dropped the name "Presidio of San Francisco" and began referring to the area as "Idle Acres."[34]

Despite pressure to dispose of the Presidio, the Army insisted that the post was essential to national defense, although some critics likened its strategic importance to that of the Alamo in the Southwest and Yorktown in the East. One colonel circulated a document arguing that the Presidio, if fortified, could offer protection against thermonuclear attack.[35] It is not surprising that Army officials felt so strongly about the role of the Presidio. After all, it had been a military post for a century, during which time it had earned the reputation as one of the best duty stations in the Army. Its golf course, bowling alley, and movie theater, Baker Beach, and spectacular views of the city and the Pacific Ocean made assignment to the Presidio one of the most highly sought-after tours of duty in the Army. Once again, the post needed to emphasize its own importance, because its future as an active military post was in jeopardy.

Army efforts to keep the Presidio as Sixth Army headquarters galvanized some local citizens to form a group to save the Presidio. In 1957, several residents founded a nonprofit corporation, the Presidio Society, to support the historic role of the Presidio and to aid in its preservation, improvement, and beautification. In 1962, an attempt to use part of the Presidio for development was derailed by citizens of San Francisco, in part by lobbying the Department of the Interior for the Presidio to be designated a national historic landmark. It was hoped that such a designation would thwart development of certain areas and protect historic buildings. In 1963, the National Park Service designated the Presidio a national historic landmark, citing the Mission Revival buildings of Fort Scott as particularly noteworthy. At the dedication ceremony, Secretary of the Army Stephen Ailes reassured the crowd that the Presidio would remain part of the Army.[36] Still, between 1945 and 1970 Army activity at the Presidio declined noticeably, while the campaigns for real estate development at the Presidio gained strength.[37] Advocates for preserving the Presidio—green, beautiful, and military—struggled to keep the developers at bay.

Pro Growth, Slow Growth: Reactions to Urban Development

Meanwhile, in the postwar years of American economic prosperity, the city and peninsula saw expanding growth and residential development. Numerous industrial and business parks, residential development, schools, and shopping centers were built across the once rural landscape. Many of the thousands of new military personnel who had been stationed in the Bay Area during the war settled permanently, multiplying the need for housing, schools, and shopping areas. San Francisco and other Bay Area cities and communities, already swollen with the influx of war industry migration, expanded outward in all directions, into city outskirts, fields, and orchards. In the postwar years large corporations such as Wells Fargo Bank, Levi Strauss, Del Monte, Crocker Bank, and IBM located either their corporate or regional headquarters in the city's central business district. The explosion in business growth gave the downtown a new and dramatic profile through a building boom in skyscrapers. Between 1966 and 1971, thirty-one skyscrapers were constructed and office space increased by nearly 50 percent.[38] As more and more high-rise buildings were erected, they often obscured the views of the bay and the hills. This trend in high-rise development generated public debate. Not everyone welcomed skyscrapers as towering monuments to technology and economic prosperity. Herb Caen, the widely popular writer and the city's unofficial biographer, lamented what he called the "Vertical Earthquake" of downtown development that had succeeded in tossing up "sterile stacks of ice cube trays in the sky."[39] Caen worried that postwar San Francisco lacked style and feared that "soon all will be new, bright, shiny and soulless." Sentiments such as these grew more common as residents began to complain that San Francisco was being Manhattanized into a series of dense residential and office buildings.

Growth in San Francisco and the rest of the Bay Area also proceeded horizontally. Between 1945 and 1957, state highway engineers completed more than 200 miles of freeways to link the growing metropolitan region; central business districts expanded in cities all over the peninsula; and a boom in the construction of office buildings made skyscrapers a familiar sight. Redevelopment projects in the 1950s and 1960s converted old produce markets and industrial spaces into administrative and financial institutions; neon signs and billboards went up in all directions. The lack of zoning or inappropriate zoning along with the relentless pressure from developers resulted in detrimental uses of land.[40] Residential building took place on hillcrests and mountains, creeping into nooks and crannies. Apricot and pear orchards were obliterated to make room for houses and business complexes. Urban development even began to engulf the open spaces, and the likelihood of continued development further threatened their loss. In the 1960s, Bay Area counties were losing 25 square miles of open space each year.[41]

The combination of vertical growth downtown and horizontal expansion in the outlying communities prompted considerable public discussion about the protection of open space. In the postwar years, a "progrowth regime" had dominated urban politics and sold San Francisco as an emerging transpacific urban community.[42] But critics argued that development was chaotic, uncoordinated, and out of control. Skyscrapers and uncontrolled development had left a paucity of open public space in central areas. Critics chastised downtown development for ignoring the natural topography of the city's landscape, casting shadows in the downtown area, and creating powerful wind tunnels. In response to the destruction of the skyline and other negative consequences of rapid growth in the Bay Area, a "slow-growth movement" began to mobilize. This coalition wanted to combat progrowth development by prioritizing the needs of local residents rather than those of outside developers and investors, and also by emphasizing the city's human and natural use values rather than its purely economic or exchange value.[43] Out of the concern about the path of postwar development, several interest groups and social movements formed, using the political tools of democracy such as initiatives and referendums to combat development projects. Neighborhood coalitions were successful in revolting against plans for freeways and housing projects. Citizens formed a task force, California Tomorrow, which published a book outlining an alternative to the haphazard postwar development.[44]

The Emerging Environmental Movement

By the late 1960s, concern over environmental quality and pollution was coalescing into a powerful social movement. Industrial and municipal wastes flowing into the bay had created a vast sinkhole so contaminated that many waterfront areas were closed to the public.[45] Air pollution from automobiles and industrial manufacturing, and water pollution from industrial dumping and sewage disposal, erupted as important problems. Concern about the insufficiency of parks, especially the lack of a regional park system, grew. In San Francisco, well-informed, highly educated, and vocal citizen activists fought for new legislation to reduce pollution and to enlarge recreational areas and preserve Bay Area scenic assets. Urban development and concern about environmental quality helped to focus citizen concern for Bay Area ecology.

Attention to environmental issues mobilized the formation of local environmental groups that focused on enhancing the quality of life through regulation and the control of growth. At a fundamental level, these groups attempted to redefine urban form, function, and meaning. The meaning and importance of certain places in the city are results of ongoing processes of challenge, debate, retreat, and compromise, and urban social movements—such as slow-growth movements, tenants' struggles, a neighborhood association, or an environmental organization—can challenge prevailing cultural and economic values. In the 1970s, Bay Area environmentalists challenged prevailing values in two ways: first, by questioning the direction of urban development (challenging current or proposed land use); second, by lobbying for the protection of the natural environment (challenging the practice of land abuse).

In the 1960s and 1970s, several citizen groups formed to promote environmental protection. Citizens for Regional Recreation and Parks, later renamed People for Open Space, worked for the setting aside of land for public parks. Citizen activism by this group helped pressure the National Park Service to purchase properties for the 53,000-acre Point Reyes National Seashore in Marin County in 1962.[46] The Point Reyes project was a blend of efforts to preserve natural heritage, an ecologically unusual landscape, and widely visited public open space. People for Open Space and various grass-roots groups in Bay Area locales continued to influence park acquisitions and protection of open spaces. For the next thirty years, thousands of parks were added in towns, communities, cities, and counties around the Bay Area. The protection of open spaces and enlargement of parks seemed to become important new values, evidenced by the fact that people voted again and again to increase their taxes in order to acquire new parklands.[47]

Citizens also formed the Save the San Francisco Bay Association to protect the bay against plans to reclaim or fill in selected marsh areas in order to accommodate urban development. They were successful in pressuring lawmakers to create the Bay Conservation and Development Commission in 1965. This organization was charged with creating a comprehensive plan for the future of the bay and providing continued public access to the bay while allowing urban development projects to continue.

In addition to many local grass-roots organizations, San Francisco became home to several national nongovernmental environmental organizations, including the powerful Sierra Club, Friends of the Earth, and Greenpeace. The Sierra Club board of directors charged local chapters to embark on programs to preserve and protect parks at the city and regional level. In San Francisco and around the Bay Area, local Sierra Club chapters began to envision an interconnecting system of parks. The environmental movement, which emerged as a *national* social movement in the early 1970s, was present as a powerful *local/regional* movement in San Francisco a decade earlier. The city was at the forefront of what would become an enduring social movement. The combination of an antigrowth coalition and the emerging environmental movement was to have a profound impact on the Presidio.

The Presidio as Park: The Second Wave of the Urban Park Movement

The urban park movement of the 1860s through 1880s reflected new ideas about the value of parks and urban nature, changing the way city dwellers related to the natural world and leaving a legacy of parks in many cities around the country. One hundred years later, in 1970, the modern urban park movement reemerged, tied to two developing political coalitions: the antigrowth movement and the national environmental movement, which was becoming more influential in mainstream American politics.

In 1970, as the environmental movement crystallized, President Nixon's secretary of the interior, Walter Hickel, promoted a new idea to create *urban* parks as part of the national park system. The new parks would be called national recreation areas. Hickel explained: "We are moving with a coordinated program to establish large parks and recreation areas where most of our people live—in the metropolitan areas of our country. In past years there has not been sufficient federal emphasis on providing funds for recreation and open space preservation in and around our large cities where we believe the needs are greatest."[48] Hickel had the support of several powerful members of Congress, including California senator Alan Cranston, who noted that

> only a relatively small number of Americans have the opportunity to enjoy the wide range of natural wonders [the National Park System] protects and preserves. Those fortunate enough to visit distant units of the National Park System are most likely white, educated, relatively well-off economically, young, and suburban . . . I believe that we have a responsibility to bring the parks to the people, especially to the residents of the inner-city who have had virtually no opportunity to enjoy the marvelous and varied recreation benefits of our national parks.[49]

Hickel's proposal, made during an election year, bolstered Nixon's poor environmental record. In 1971 in his State of the Union message President

Nixon announced his administration's "Parks to the People—Where the People Are" program. This new program would seek to preserve open spaces in and around cities, because in many cases the cost of acquiring and maintaining such open spaces greatly exceeded the funding ability of local government. Responsibility for the new urban parks would reside with the National Park Service.

At the time Parks to the People was proposed, several prodevelopment groups were again suggesting new residential and private development projects for the Presidio. Citizens rallied together to protest the threat of development. Concern about the fate of the Presidio, and the possibility that it would be subdivided and developed as high-rise condominiums by private real estate interests, mobilized citizen action. Local citizens proposed a Bay Area park system for inclusion in the National Park Service's Parks to the People project.

Two influential and involved San Francisco residents, Edgar Wayburn and Amy Meyer, were among the founders of the organization People for the Golden Gate National Recreation Area (PFGGNRA, pronounced, or so it has been alleged, "Piffgunura"). Wayburn, a passionate conservationist and five-time Sierra Club president, led the crusade.[50] Wayburn and Meyer then approached local congressman Philip Burton to enlist his support for protecting various open spaces around or in the city from development. Not known as an environmentalist, Burton was, however, an astute politician who would never pass up the opportunity for a public affairs coup. He suggested that Wayburn and Meyer draft a proposal to create the Golden Gate National Recreation Area, telling them to be more ambitious than their initial proposal suggested. Burton realized that the Presidio had outlived its military usefulness and its closure was a political inevitability; he insisted that Wayburn and Meyer include the post in the proposed Golden Gate National Recreation Area.[51]

The draft legislation specified that the Golden Gate National Recreation Area (GGNRA) would encompass the shoreline areas of San Francisco, Marin, and San Mateo counties, which contained a variety of natural environments—ocean beaches, lagoons, marshes, and meadows. In total, the GGNRA system would include Alcatraz, Army lands on both sides of the Golden Gate, the Presidio and its historic landmarks, Ocean Beach, Stinson Beach, Muir Woods, various city parks including Aquatic Park, the Cliff House, the San Francisco National Maritime Museum, and much of the Marin coast up to and including the Point Reyes National Seashore (see Figure 4.6). In addition, the draft legislation provided for the establishment of a

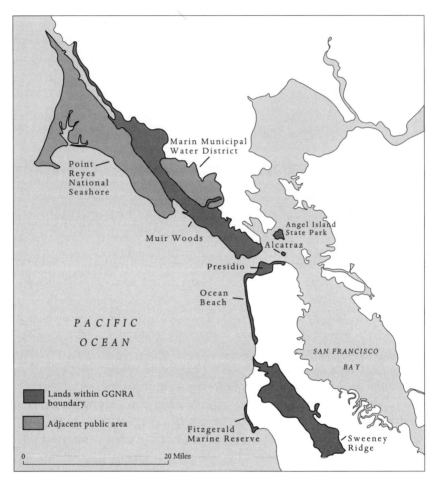

Figure 4.6
The Golden Gate National Recreation Area, 1990. *Source: Map by author, adapted from a National Park Service map for the Golden Gate National Recreation Area, 1990.*

nonprofit citizens' advisory committee, the Golden Gate National Park Association, to serve as a vehicle through which the community would be included in planning and managing the GGNRA. In all, the proposed GGNRA stretched along 28 miles of coastline—the largest undeveloped urban shoreline in the nation. And, although the park areas were not contiguous, unity of purpose linked them: protection from development and preservation for urban dwellers and future generations.

The proposed GGNRA fortuitously included a variety of "surplus public lands." Indeed, most of the large open spaces covered by the proposal were the result of the undeveloped state of federally owned lands located on both sides of the Golden Gate. Included in the proposed legislation was a brief paragraph in Section 8.i, which stated, "all lands within the Presidio of San Francisco or [that] border on the San Francisco Bay or the Pacific Ocean and that heretofore or hereafter are declared excess to the needs of the United States shall be transferred to the administrative jurisdiction of the Secretary [of the Interior]." This inclusion meant that if the Presidio was ever closed, it would be protected from development and transferred to the Department of the Interior.

Burton began his promotion of the GGNRA legislation in 1972. Luckily for him, 1972 was an election year, one in which environmental issues were playing a large role. Also, at that time there was broad bipartisan support for legislation to preserve land for public use before it was lost to development. For these reasons, and because of Burton's shrewd political maneuvering for congressional support, President Nixon pledged to sign the bill if Congress passed it.

In the House and Senate hearings on the proposed legislation, it became clear that the creation of the park system had wide support. Political elites and more than twenty-five local organizations testified. Leo McCarthy, a San Francisco assemblyman, commented that "The GGNRA is a concept so clearly in the local, state and national interest that not even a small murmur of dissent has been heard."[52] Support was overwhelming and widespread.[53]

A crucial element in the legislation, and the major point of debate in the hearings, was the inclusion of Presidio lands in the GGNRA. Burton had included the entire post, not just the coastal areas little used by the Army. This inclusion challenged Army jurisdiction at the Presidio and generated no small measure of controversy. The Army argued against the inclusion of the entire Presidio, realizing that it would lose control of any development on the Presidio—even small military projects would be subjected to Department of Interior approval. Impassioned debate followed. A PFGGNRA statement argued for the inclusion of the Presidio:

> We want the green, open spaces of the Presidio to stay just as they are, green, open, and not built upon. . . . The best way to protect the Presidio's wondrous open spaces from further building is to transfer those areas to the jurisdiction of the National Park Service of the Department of the Interior. If the Presidio is to retain its park-like character, it is essential to include it into GGNRA.[54]

McCarthy concurred: "the GGNRA is a doorway to a new kind of open space, and the GGNRA hinges on the Presidio . . . without most of the Presidio lands to anchor it—lands admittedly of no military use—the GGNRA's main concept is destroyed."[55] Despite the Army's record of conservation and beautification, many locals felt there was a continued threat of military or federal development of the Presidio that could undo the century of Army conservation. San Francisco Board of Supervisors member Ronald Pelosi commented that "while the City is grateful to the Army for maintaining the large open space areas of the Presidio and for making these available for public enjoyment, we do want to assure the continued open space character of these lands and their improvement for purposes of recreation and historic interest. Under the Department of the Interior, we have such guarantees."[56]

Although the concept of appropriating surplus military land was not new—Alcatraz, Angel Island, and parts of Fort Baker had been declared "in excess" to military needs in the 1960s—no one had appropriated military land not yet declared surplus. And few people were predicting that the Presidio lands would become surplus anytime soon. Rather, the Presidio's inclusion in the GGNRA legislation represented a long-term attempt to protect it against urban development. Congressman Burton supported the inclusion of the Presidio by explaining,

> Fortunate circumstances have saved this area from the encroachment of the urban metropolis beside it. However, this good fortune cannot be depended upon to continue. Revised military needs and urban population pressures threaten these acres of open lands. If we in the Congress do not act, the majestic area where sea and bay and land meet in a glorious symphony of nature will be doomed.[57]

Indeed many feared that because of the Presidio's location, Presidio lands, even if deemed surplus by the military, could be appropriated by other federal agencies. These fears were not unfounded. The Food and Drug Administration had already outlined plans to develop a research program at the Presidio. Burton and others realized that without a strictly defined and inclusive plan to control development of the Presidio as an entity within the park system, other federal agencies could appropriate the post.

If the entire Presidio were under Interior Department jurisdiction, any federal agency (including the Army) would be required to meet National Park Service guidelines and obtain approval for any development project.[58]

This became known as the "One Up, One Down Provision." For any new development the Army proposed at the Presidio, the Army was required to remove an equal amount of building space. This meant the Army would have to keep an inventory of the amount of space it was constructing and identify specific buildings that it would remove in order to have the right to build a new facility. In addition, sale of any Park Service lands would require congressional approval, thereby protecting Presidio lands (and the rest of the GGNRA) from being sold to private developers. The inclusion of the entire Presidio into the proposed GGNRA and the "One Up, One Down Provision" gave ultimate control of any future projects at the Presidio to the National Park Service, not the Army. Burton's legislation thus avoided putting the Army in the park business, but it also gave the Department of the Interior ultimate control over this space. In doing so, it redefined the entire long-term form, function, and meaning of the Presidio from that of a military post to that of a park/recreation area. It was a significant piece of legislation.

Burton was successful in persuading Congress to support legislation that included the entire Presidio within the GGNRA. In 1972, Congress established the Golden Gate National Recreation Area (GGNRA), an area of some 27,000 acres, under the National Park Service, and President Nixon signed the bill into law on October 28, 1972.[59] The GGNRA was one of the nation's first urban national parks, established "in order to preserve for public use and enjoyment certain areas of Marin and San Francisco Counties, many of which are surplus public lands."[60] The final draft of the GGNRA legislation stipulated that "when the Department of Defense determines that the Presidio lands were determined to be excess to its needs, such lands would be transferred to the Department of the Interior for national park purposes." This small inclusion would have far-reaching impacts on the future of the Presidio.

The significance of this achievement should not be underrated. Without Burton's foresight, the Presidio and many of the other areas now protected as GGNRA land might have been put on the auction block. There are two ways to look at the creation of the GGNRA. First, someone had to develop a plan to protect trees and meadows in a place where others were dreaming of high-rise condominiums and office buildings. Second, there had to be "a campaign to transform this dream into law, a process that entailed creating constituencies from thin air, sweet-talking bureaucrats, lobbying Congress and generally pulling rabbits out of hats."[61] Within three years, local citizens had helped to create one of the finest and most extensive urban parks in the United States. Over the course of the next twenty-three years, the GGNPA and the National Park Service would position the Presidio as an integral part of the long-term vision of the GGNRA.

In 1989, the Golden Gate National Recreation Area encompassed more than 70,000 acres of open space (of which 29,000 are federally owned). Several boundary revisions and property acquisitions over the years had added greatly to the acreage. Since its inclusion in the GGNRA, the Presidio had become a destination for many city dwellers looking for inexpensive daytrips to parks and open spaces—places that were accessible to the public in the face of overcrowding at many Bay Area parks.

In 1989, the Presidio encompassed approximately 1,480 acres. The post contained more than 60 miles of roads and hiking trails. There were more than 800 buildings making up an estimated 6.3 million square feet of building space. It was, most certainly, a multiuse space. In addition to the Sixth Army, several nonmilitary tenants occupied the post, including the Presidio Golf Course, the Public Health Service Hospital, and concessionaires such as Burger King. In the years since the establishment of the GGNRA, the National Park Service had assumed responsibility for several of the deactivated areas of the Presidio, including Fort Point and Baker Beach. It also established guided tours, nature hikes, and school programs for local children. Recreational opportunities at the Presidio included hiking, horse and bike trails, beach access, and picnic areas in meadows and on the edge of ocean bluffs. Open space areas included the shoreline of Crissy Field, the Presidio Hill (including its golf course), the coastal bluffs, Baker Beach, and the Presidio Forest, a 400-acre forest of mature eucalyptus and evergreen trees. Naturalists could study the Presidio's several plants that were on the federal list of endangered species. Among these were the Presidio (Raven's) manzanita, a rare shrub; the Presidio clarkia, a pink wildflower; San Francisco popcorn flowers, and the San Francisco lessingra. And every year, more than 100,000 people crowded the beaches and promenades of Crissy Field to celebrate the Fourth of July. The Presidio was an assemblage of one of the most diverse collections of military architecture in the nation, ranging from the remains of an original adobe wall, to Civil War–era wooden barracks, to Victorian-era officers' quarters, to turn-of-the-century Mission Revival buildings. San Franciscans had come to view the Presidio as a well-loved park and national historic landmark rather than a government installation. Although the Presidio was considered an urban open space, it was very different from Golden Gate Park. At the heart of its legacy were two centuries of military history and architecture. The Presidio had been a witness to events that helped shape and define American identity.

It was only a matter of time before the Army would declare the entire post "military surplus space" and the Presidio would be wholly transferred to the National Park Service. Events during 1989 set in motion the Presidio's ultimate transformation.

Reflections on the Presidio's Historical Geography

The history and geography of the Presidio are important for understanding the present-day Presidio and its future as park/recreation area.

The Presidio's strategic importance as a military post has waxed and waned. During a series of wars, campaigns, and operations it regained its importance to the defense structure of the West. Concomitant with its periodic heightened importance and increased troop strength were new construction projects and development. In the long periods between wars and conflicts, its infrastructure suffered neglect, in many instances to the point of near abandonment. At times, under the command of the Spanish, Mexican, and American military, the Presidio's purpose and strategic significance were unclear. Ironically, during the years of peace, Army planners and engineers gave to the Presidio perhaps its most enduring legacy by turning from troop engagement to environmental conservation. In this respect, the Army played an unwitting but voluntary conservationist role.[62] Army planners and engineers initiated a beautification program in which troops planted more than eighty thousand trees around the post. Other improvement projects included a golf course, hiking trails, picnic areas, and scenic drives.

As one of the oldest operating military posts, the Presidio contains a diverse collection of architecture, and more than three hundred buildings on the post have been designated national historic landmarks. Federal (military) stewardship of this land proved invaluable; it protected the post from urban development and preserved it as an urban open space.

It is worth emphasizing that although the Presidio was always a military post, it also was part and parcel of a larger urban space. One of the most distinguishing attributes of the Presidio has been its enduring interdependent relationship with the city of San Francisco as a public open space/urban park. The perception of the Presidio as an "open post" reflects the often contradictory character of the place: accessible to the public yet distinctly military property. This ambiguity manifested itself most clearly in recurring attempts by various citizens or groups to annex or preserve Presidio lands as part of the city park system. Many San Franciscans thought the Presidio was *of* the city. Former San Francisco mayor Joseph Alioto illustrated this when he noted, "in tandem, the Presidio and Golden Gate Park give us as fine a park setup as any city has anywhere in the world."[63]

Thus the Presidio is different from other city parks; it also is different from most military bases. Not just a military base, or simply an urban open space, it is a place significant in its historic links to the city, its design, and its symbolism. In the hearts and minds of San Franciscans, the Presidio is a very special place. Vikram Seth celebrated the city and the Presidio in his novel in verse, *The Golden Gate* (1986):

> It's dark. He drives. The street lamps glimmer
> Through cooling air. The golden globes
> By City Hall glow, and the glimmer
> —Like sequins on black velvet robes—
> of lights shines out across the water,
> Across the bay, unruffled daughter
> Of the Pacific; on the crests
> Of hill and bridge red light congests
> The sky with rubies. Briskly blinking
> Planes—Venus-bright—traverse the sky.
> Ed drives on, hardly knowing why,
> Across the tall-spanned bridge. Unthinking,
> He parks, and looks out past the strait,
> The deep flood of the Golden Gate.
>
> Subdued and silent, he surveys it—
> The loveliest city in the world.
> No veiling words suffice to praise it,
> But if you saw it as, light-pearled,
> Fog-fingered, pinnacled, I see it
> Across the black tide, you'd agree it
> Outvied the magic of your own.
> Even tonight, as Ed, alone
> Makes out the Marina, plaza tower,
> Fort Point, Presidio—he feels
> a benediction as it steals
> Over his heart with its still power.[64]

On occasion, people form profound relationships with a place. This, says geographer Edward Relph, is as unavoidable as close relationships with other people, because places are incorporated into the intentional structures of all human consciousness and experience.[65] The essence of place comes not from its trivial functions (for example, the military purpose of the post) but from the intentions and efforts (self-conscious or not) that define the place as a center of human experience.[66] This is why the Presidio, as a cultural landscape, has such significance for residents of San Francisco.

In the past, some geographers have used the term *sequent occupance* to describe the way in which human activity transforms a place over time.[67] Human occupation of an area results in layers of transformation in the economy, culture, and physical landscape. Although this concept is not as widely used today, its premise is useful for seeing the history of the Presidio as a succession of stages of human occupancy and their cultural imprint—from the Ohlone, to the Spanish, to the Mexican, to the Americans. Each generation has also transformed the Presidio—from the Spanish era of missions and cattle ranches, to the early gold and silver economy of San Francisco, to the rapid urbanization of the twentieth century. The history of the Presidio is a product of the different ways former inhabitants used the land; the present is clarified by pointing out the impact of the past.

The present-day Presidio is a product of human design over many years by many people. It has become a repository of symbols and experiences, all of which are layered with meaning and significance. Over its long history, many people have cared about the Presidio and have attempted to improve or redefine its infrastructure, its architecture, and its natural landscape. From the earliest Spanish and Mexican comandantes who requested additional money for repair and supply, to Army officers who redefined the post's strategic defense purpose, to the Army planners who recognized the special relationship between the post and the city and implemented a forestation program to beautify the post, to officers who convinced their superiors that architecture at the Presidio should reflect vernacular rather than standard styles, to the many citizens and citizen groups who mobilized to protect the Presidio from development, to the congressman who, by securing legislation, helped ensure the Presidio's unique future as part of a park system. These are but a few examples of some of the major agents who helped to construct the Presidio landscape. This "defining force" of human intention and historical events has made the Presidio an important place, not only in military terms but also in cultural terms.

The Presidio is perhaps unique among military bases in that it was simultaneously a city within a city and a park within a city. Roger Kennedy, director of the National Park Service, gives perhaps the most eloquent of words to this concept: "The Presidio is a community within a park within a larger community. We are reminded by such accidents of geography that each of us is placed in human life within the concentric circles of relationship to others and to the natural world."[68] This concentricity owes a great deal to its intimate and intertwined relationship with the city, but its natural and cultural attributes are not a geographical accident. They are the outcome of a series of local urban and park-planning decisions, national social movements such as the urban park movements of the nineteenth and twentieth centuries, and broader international events, such as wars and conflicts.

The proposed future of the Presidio as an urban park is not unique to recent urban planning or unique to the twenty-six-year-old GGNRA. Rather, the Presidio as park has been a recurring vision for centuries. Historically, there have been many efforts to enhance this vision of an urban park: the history of the Presidio as an open post; the urban park movement of the 1880s, which resulted in the beautification and forestation of the post; increased public access resulting from the construction of the Golden Gate Bridge and coastal scenic drives; and the second wave of the urban park movement in the 1960s and 1970s, which resulted in federal legislation protecting the Presidio from future development. The concept of Presidio as national park is not a revolution in planning ideas or a radical redefinition of space but rather the century-long evolution of a vision.

CHAPTER 5

Post–Cold War Military Base Closures: From Swords to Plowshares

In 1988, Congress passed the Defense Base Realignment and Closure Act. This act established a bipartisan commission to decide which military bases were obsolete and no longer necessary to the defense of the United States. Since 1989, the commission has designated more than five hundred military installations and facilities around the country for closure or realignment, the most extensive such action in U.S. history.

Although the Presidio was only one among five hundred military bases or installations due to be closed or realigned, its future as part of the Park Service's Golden Gate National Recreation Area made it unique among military bases. But the way in which the post came to be declared "surplus military property" was not unique. Rather, the story of the Presidio from 1988 to 1995 is, partly, connected to the larger story of changes to the U.S. military structure.

Historical Cycles of Defense Spending

> They shall beat their swords into plowshares, and their spears into pruninghooks: nation shall not lift up sword against nation, neither shall they learn war any more.—Isaiah 2:4[1]

At the conclusion of each war in U.S. history, Americans have beaten their swords into plowshares—through defense reductions, demobilization, and restructuring.[2] In the last half of the twentieth century, cycles of defense-spending reductions followed the end of World War II, the Korean War, the

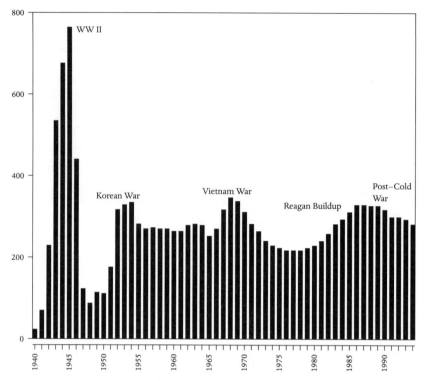

Figure 5.1
Cycles of Real Defense Spending and Reduction, 1940–1994 (in billions of 1991 dollars). *Source: Adapted by author from data in U.S. Congress, Office of Technology Assessment, After the Cold War: Living with Lower Defense Spending (Washington, D.C.: GPO, February 1992).*

Vietnam conflict, and the Cold War (Figure 5.1 illustrates these cycles of defense reductions). Defense reductions or drawdowns have entailed job losses for both military personnel and Defense Department civilian employees, the reduction or elimination of private-sector defense contracts and hence local jobs, loss of federal investment at the local level, and short-term economic distress, often in the larger urban and regional economies.

Defense reductions after the Second World War were the most dramatic, particularly the personnel reductions. Although large-scale demobilization characterized the years following 1945, the changeover from military to civilian work was accomplished without drastic economic distress, in part because the large industrial firms that did war work until 1945 went back to producing the commercial items they had made before the war and with which they were completely familiar.[3] The changeover after the war was a process of reconversion rather than a process of conversion and was assisted by the growing demand for consumer goods.[4] Despite extensive defense reduction and restructuring, this adjustment to the peacetime economy was made without apparent difficulty and was followed by a period of economic prosperity.

Defense reductions after the Korean War, between 1953 and 1956, were smaller in scope but generally caused more economic distress. The economy went into recession in 1954, recovered, but declined again in 1958. A Defense Department study argued that although defense reductions resulted in short-term pressure on the national economy in all cases, after several years the economy recovered robust growth.[5] Defense reductions in the post–Vietnam era were affected by two factors: first, military disengagement from Vietnam; second, Defense Department streamlining efforts that were independent from post–Vietnam reductions and instead related to inflation-adjusted defense spending.[6] This phase of defense reductions, characterized by significant military base closures and realignments, is more temporally drawn out.

Concomitant with ongoing military base closures (from 1961 to 1968 and from 1973 to 1974) was a more pronounced series of defense-spending and personnel reductions (from 1968 to 1974) that followed combat disengagement from Vietnam. It is worth briefly discussing these streamlining efforts because they highlight the important fact that the Defense Department has engaged in a process of assessing the military needs of the nation and continually reviewing the military base structure in light of changing requirements. The concept of closing or realigning military bases is no exception; it has been an integral part of how the department views defense organization and structure.

In 1961, during the Kennedy administration, Secretary of Defense Robert McNamara initiated a study of bases that concluded that many were superfluous and could be converted to civilian use. From 1961 to 1969, over 950 defense installations were identified for cutback, reorganization, or closure.[7] The Defense Department projected a cost sav-ings of nearly $1.6 billion annually.[8] At that time, it was the most extensive base closure program in the history of the military. Following these closure announcements and negative community reactions, the department "re-emphasized that the Pentagon was responsible for the security of the nation and not really for the creation of a level of demand adequate to keep the national economy healthy and growing."[9] Reinforcing the rationale for streamlining efforts, President Johnson said in 1964, "We expect to receive a dollar value for every dollar we spend on defense. To do so has meant we must eliminate surplus, obsolete and unnecessary defense installations."[10]

In 1961, Congress created the Office of Economic Adjustment (OEA) to help communities plan for base closings. OEA was empowered to offer grants to help plan for defense conversion and to develop long-term strategies for stimulating local economic growth. The involvement of OEA in assisting many communities with military base closures helped to popularize the terms *economic adjustment* and *economic conversion* to describe the challenges of promoting economic development and diversity in local communities after the disruptions caused by changes in defense spending.[11] *Economic conversion* came to be defined as "a process which includes the formulation, planning and execution of organizational, technical, occupational and economic changes required to turn manufacturing industry, laboratories, training institutions, military bases and other facilities from military to civilian use."[12] Since the closures and realignments of the 1960s and 1970s, various researchers have attempted to analyze community efforts at economic conversion and to provide models of successful conversion projects.[13]

THE SPATIAL CONSEQUENCES OF DEFENSE INVESTMENT, 1945–1989, AND THE RISE OF THE GUNBELT

Reductions in defense spending and troop demobilization were quite dramatic after the end of the Second World War (see again Figure 5.1). But while the federal government was reducing military spending from wartime peaks, many politicians embraced a form of Keynesian economics that situated military spending as a regulator of market demand and as a stimulator of economic growth in U.S. industry.[14] In addition, the United States moved from fighting a world war to waging the Cold War. The emphasis shifted from military engagement to military deterrence that relied on technological innovation rather than on combat forces. This shift to research and development created what President Eisenhower in 1961 called the "military-industrial complex," an institutionalized, interdependent partnership among the armed forces, private corporations, universities, and the federal government, which provided the nation with the technologies of war. The military-industrial complex promoted the rise of private-sector defense contractors (the defense industry) and established national research laboratories to work on the technological development of sophisticated weaponry and equipment, such as computers and nuclear weapons. The creation of the military-industrial complex has led Seymour Melman to contend that "since 1951, the

DoD has been made into the largest single controller of finance capital in the American economy."[15] Although dramatic demobilization of troops and personnel followed World War II, the Cold War rationale and the growth of the military-industrial complex during the 1950s and 1960s made Defense Department needs a significant part of the federal budget.

The establishment of the military-industrial complex had both economic and spatial consequences. Military-related economic growth created new military communities and was a crucial factor in the remapping of industrial America.[16] These new military communities have been concentrated unevenly around the U.S. coastline, geographically arranged in a peripheral belt that runs from Boston to Tampa in the East, from Huntsville to Tucson in the South, and from San Diego to Seattle-Tacoma in the Pacific West. Geographers call this high-growth, high-tech, defense-related periphery the "Gunbelt" (see Figures 5.2 and 5.3).[17] The primary Gunbelt states are Cali-

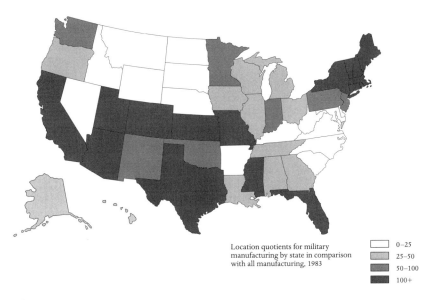

Location quotients for military manufacturing by state in comparison with all manufacturing, 1983

- 0–25
- 25–50
- 50–100
- 100+

Figure 5.2
The Gunbelt. "Superimposed on the map of America, these new military-industrial regions do indeed resemble the belt around the hips of the solitary sheriff in pursuit of the bad guys in an old western movie. The Southwestern states, Texas and the Great Plains make up the holster; Florida represents the handcuffs . . .; New England is the bullet clip." *Source: Adapted from Ann Markusen, Peter Hall, Scott Campbell, and Sabina Deitrick,* The Rise of the Gunbelt: The Military Remapping of Industrial America *(New York: Oxford University Press, 1991), 3–4.*

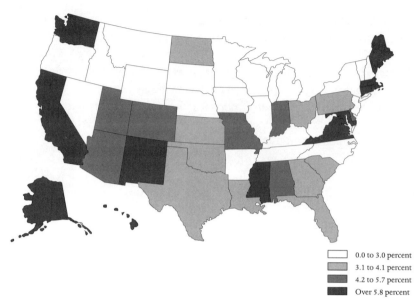

Figure 5.3
Defense Spending as a Percentage of State Purchases, 1991. *Source: U.S. Congress, Office of Technology Assessment,* After the Cold War: Living with Lower Defense Spending *(Washington, D.C.: GPO, February 1992: 13).*

fornia, Connecticut, Massachusetts, Missouri, Texas, and Virginia. Military spending has thus had a profound impact on local economies and on urban and regional development, and some of the regional inequity of the past twenty-five years can be attributed to Defense Department investment and the expansion of the defense industry.[18]

Two historical factors have been affecting defense spending since 1945. The first has been the cyclical nature of defense spending and reductions that have followed war and conflict. After each war since World War II, there have been concentrated efforts at defense reductions. After World War II and Korea, defense reductions focused on troop demobilization and personnel reductions; during the Vietnam era, defense reductions focused first on the defense support structure—specifically, on military base closures and realignments—and then on troop demobilization. Thus defense reductions in the aftermath of the Cold War have been consistent with historical precedents. In addition, we might expect defense reductions in the post–Cold War era (a war of deterrence rather than engagement) to focus on military base closures and realignments rather than troop demobilization. Indeed, the Defense Department has targeted defense support structures—military bases and facilities—as the linchpins in its reduction efforts.

The second historical factor at work has been the spatial impact associated with significant federal investment in defense and private-sector defense contractors, despite the cyclical patterns of defense spending. The Gunbelt, for example, emerged as a result of continuing federal investment in certain cities, states, and regions over the past fifty years. The concept of the Gunbelt has been useful in displaying the geographic patterns of economic development in certain states and regions based on federal defense investment.

Given that defense reductions and military base closures are not new or exceptional, what might explain the perception of current military base closures? The dominant interpretation of base closures is that they are a "post–Cold War peace dividend."[19] This is partially true. There is, however, another explanation for these defense reductions. Economic conditions during the 1980s resulted in congressional efforts to reduce defense spending.

BOLSTER THE HOLSTER: ECONOMIC FACTORS AND THE DEFENSE BUDGET

If the 1970s were characterized by détente, military scale-downs, and base closures, the 1980s were characterized by an escalation of Cold War rhetoric and very significant defense buildups. Between 1976 and 1985, defense appropriations nearly tripled, increasing from $89.6 to $253.8 billion.[20] During the 1980 presidential campaign, Ronald Reagan promised to rebuild a neglected military and to reassert America's military muscle around the world. Reagan's "Make America Great Again" theme "encompassed a broad range of security issues, including calls to modernize the nation's military forces, roll back Soviet political gains in the Third World, and restore international respect for American power and authority."[21] Military reinvigoration stressed technological advantage and the mobility of military forces. As a result of Reagan's military program, the 1980s were devoted to a defense buildup unparalleled during peacetime. According to the Center for Defense Information, from 1980 to 1988, the Reagan administration spent more than $2 trillion on defense[22] and appropriated approximately $23 billion to the development of the Strategic Defense Initiative (known informally as "Star Wars"). The Pentagon was awash in cash and contracts, and defense contractors enjoyed large profits.

Reagan's emphasis on technological advantage and modernization encouraged U.S. manufacturers to increase their involvement with Defense Department research and development. Military-oriented manufacturing jobs flourished; the number of firms with more than 10 percent of their output devoted to defense orders rose from twenty-one in 1977 to forty-five in 1985.[23] The military buildup of the Reagan years was a major contributor to American job growth in the 1980s, accounting for at least a 17 percent increase in overall job growth.[24] The spatial consequences of Reagan's efforts to "bolster the holster" included continued economic investment to the Gunbelt region, perhaps at the expense of the traditional mass-production factories of the Rustbelt.[25] During the 1980s, much of the Office of Economic Adjustment's work, ironically, was helping communities adjust to Defense Department *expansion*.

By 1986, however, the Reagan administration's generous spending on defense buildup had begun to take its toll: the federal budget deficit had increased dramatically and stood at $220.7 billion.[26] Congress acknowledged it was time to address the budget deficit and reprioritize federal spending, and it pressured the administration to curb military spending. Arguments in favor of defense reductions were reinforced by a spotlight on Defense Department scandals involving exorbitant costs and corruption. In fact, 1985 proved to be the defense budget peak of the 1980s (in terms of real dollars).

The link between defense spending and economic prosperity (or economic decline) has long been a subject of dispute.[27] But by the late 1980s, it was increasingly clear that heavy defense spending was having negative consequences on a number of fronts. In 1986, defense layoffs occurred across the country (and the Gunbelt) as defense industry giants such as Rockwell International and General Dynamics laid off thousands of engineers and technicians. In 1987, Reagan and Soviet General Secretary Mikhail Gorbachev signed the Intermediate Nuclear Forces agreement, symbolizing the growing commitment of political leaders to cut national defense spending. This was the first time since World War II that governing elites agreed on the need for military cuts *in order to cope with domestic economic problems*.[28] In 1988, Reagan submitted his budget for fiscal year 1989, projecting a $129.5 billion deficit.[29] In his budget proposal, he requested nearly $299.5 billion for defense; however, this appropriation represented the smallest increase of his tenure.[30] During his annual budget summit with bipartisan congressional leaders, Reagan had been forced to trim nearly $33 billion from the $332.5 billion defense budget that originally had been planned.[31] Still, this budget expenditure represented 27 percent of federal government spending. Defense remained the second largest budget outlay, behind small benefits to individuals as Social Security and other entitlements; ironically, the third largest budget item was the provision for paying the interest on the national debt. This led one prominent businessman to comment that "part of his [Reagan's] legacy will be this deficit and this debt."[32]

By late 1988, the national debt was approaching $2.8 trillion.[33] The budget deficit had continued to grow, although the Gramm-Rudman-Hollings law mandated a balanced budget by 1994. Grappling with the massive budget deficit was a task that would dominate George Bush's early days in office.[34] Continued financial stresses on an increasingly burdened budget included high levels of household and corporate indebtedness, soaring costs of health care, commercial overbuilding, corporate restructuring, and the troubles plaguing numerous savings and loans institutions (which at the time were projected to cost taxpayers $100 billion). It was also increasingly clear that Bush would not be able to continue the defense buildup of the Reagan years. Indeed, the opposite emerged as a possibility: *reduced* defense budgets.

The Pentagon began looking for ways to cut spending without damaging readiness or canceling weapons programs launched during the Reagan era. Recognizing impending changes in budget allocations for defense spending, Secretary of Defense Frank Carlucci proposed that a base-closing commission be appointed to consider which bases were no longer necessary for national security and the mission of the Defense Department. This 1988 announcement came some twelve months before the dramatic events in Eastern Europe and eighteen months before George Bush would declare the Cold War over. Congressman Dick Armey, R-Texas, commented on the increased economic and political pressures to reevaluate the budget:

> We must look in the military budget for areas where we can make responsible savings. . . . it seems to me we have at this time an opportunity to make a contribution to the process of trimming the budget and cutting the fat out of the military budget. Saving money by closing bases or realignment of unnecessary military bases may be a difficult task, but it is one whose time has come.[35]

Efforts to combat the spiraling federal deficit were made in two ways. First, the annual defense budget could not continue to increase at the same rate as it had during the Reagan years; one solution was to reduce defense budget increases. Second, the Defense Department and Congress actively sought ways to reduce defense expenditures, essentially creating defense savings that would best utilize the shrinking budget. The second approach led to the creation of the Defense Base Closure and Realignment Commission in 1988.

Again, it is worth repeating that Congress passed legislation to appoint the base closure commission before any of the dramatic political changes in Eastern Europe and more than a year before the Cold War was declared officially over. This fact seems to indicate that military base closures were initially driven by domestic economic concerns. The end of the Cold War, however, did play a major role in the rethinking of the long-term U.S. defense structure. And, perhaps more important, it not only reinforced plans already under way to reduce defense spending but also amplified the scope of military base closures.

Coming in from the Cold (War)

The year 1989 began with the inauguration of George Bush as the forty-first president of the United States. Soon after, Hungary's parliament voted to allow independent political parties, thus initiating revolutionary electoral reform. In early February, the Soviet Union pulled its troops out of Afghanistan; in Poland, reformists demonstrated in the streets of Warsaw, spurring negotiations among the government, Solidarity, and the Catholic church. By spring, many people inside Iron Curtain countries had participated in public demonstrations and rallies. Calls for reform, freedom, and democracy precipitated vast political changes. In early June, Solidarity scored an overwhelming victory in parliamentary elections in Poland, ousting the Communist party. One week later, more than 150,000 Chinese students, who had been protesting for seven weeks in Tiananmen Square, were attacked by government forces, and many of them were killed. The photograph of the lone man who faced down a column of tanks in the street became one of the most potent symbols of protest of the year. In midsummer, protesters staged hunger strikes, rallies, and public demonstrations, all of which spread like a virus from city to city in the Warsaw Pact countries, hastening political reform. In the fall months, Czechoslovakia, Poland, Hungary, and Romania all saw dictatorships collapse or weaken and Communist parties wither. The Baltic states—Estonia, Lithuania, and Latvia—made moves toward sovereignty. The Soviet Empire was collapsing, and the Soviet Union was on the brink of disintegration.

The sudden collapse of communism culminated in East Germany in perhaps the most powerfully symbolic moment of the year. On November 9, thousands of people, armed with hammers and chisels, hacked away at the massive concrete face of the Berlin Wall. As midnight approached, the wall fell, the people cheered, and the barrier between East and West opened in a previously unimaginable stroke of creative destruction. The most potent symbol of the Cold War had come crashing down. In one bewildering instant, the Berlin Wall and everything it had so grimly stood for were simply gone. All around Eastern Europe, from Budapest to Berlin, the Communist bloc crumbled and a new world was emerging; 1989 will be remembered as the year the Cold War ended.

The end of the Cold War altered the ideological and geopolitical struggle between the Soviet Union and the United States and lessened the possibility of a military conflict between the superpowers. The Soviets' dismantling of their military infrastructure in Eastern Europe partly removed the forty-year threat of aggression, and containing communism was no longer the overriding purpose of the U.S. military. Conventional wisdom states that the dissolution of the Soviet Union and the end of the Cold War profoundly changed U.S. defense needs. Post–Cold War euphoria encouraged a rewriting of America's security agenda and a reassessment of military spending. The end of the Cold War promised to liberate billions of dollars from military budgets; resources no longer needed for national defense would be applied to long-neglected social and environmental problems. These billions were to be a reward for victory in the Cold War and were called the "peace dividend," a term that quickly entered into the lexicon of 30-second sound bites.

The Peace Dividend: A Post–Cold War Reward

On November 16, 1989, shortly after the fall of the Berlin Wall, the Bush administration's secretary of defense, Richard Cheney, announced that because the Warsaw Pact was becoming "a very different animal," the United States could reduce its defense spending and scale back the arms-race rivalry.[36] Cheney proposed further defense cuts and restructuring. Talk of realizing the elusive peace dividend began when he announced a Pentagon plan to cut $150 billion from the armed services over a five-year period (1990–1995) and to restructure the military. Even the Pentagon began referring to the base closure budget savings as a "peace dividend." National newspapers and magazines echoed talk of the peace dividend: in December 1989, the cover of *U.S. News & World Report* asked, "After the Cold War: do we need an Army?" *Newsweek* noted, "The Russians Aren't Coming," and *Time* called for "Rethinking the Red Menace," noting that "containment sounds like an anachronism . . . it is time to think seriously about retiring the NATO."[37]

The revolutionary transformations on the global stage continued and had consequences for defense policy. On January 29, 1990, Secretary Cheney, responding to continued political changes taking place in Eastern Europe and the Soviet Union, proposed further reductions in the U.S. military establishment. He also recommended that additional military bases and installations be closed. "The world has been altered in significant ways, and the result of those circumstances means that we're in a position where we can adopt a new strategy with respect to U.S. defense structure," said Cheney.[38] Cheney's plan called for a 25 percent reduction of military and civilian personnel by 1997 and corresponding reductions of facilities.[39] The collapse of the Soviet Union and heightened demands for domestic spending on social and environmental programs boosted further defense cutbacks and accelerated military base closures. Cheney explained his plan for long-term defense restructuring: "By 1995 the number of people in the U.S. military will be about one-fourth smaller than it is today. Smaller forces need fewer bases. It's as simple as that."[40]

The end of the Cold War and the removal of the Soviet Union as a major threat to U.S. national security were crucial in that they amplified planned defense reductions and restructuring efforts. Both the Bush administration and Congress agreed with Cheney's proposal for a revised defense structure. In 1991, shortly after Cheney released his plan, Congress authorized the Defense Base Closure Commission to supervise three additional rounds of base closures to take place in 1991, 1993, and 1995. Since the initial round of closures in 1988–1989, subsequent rounds have been driven by short-term Defense Department budget reductions as well as the longer-term need to restructure the nation's defense priorities as a result of geopolitical changes in Eastern Europe and the former Soviet Union.

Although base closures and defense reductions had become imperative, it is probable that the *extent* of the post–Cold War defense reductions and military base closures would have met with serious political opposition had it not been for significant legislative developments. To facilitate the Defense Department's restructuring plan, Congress had to address the very political nature of decisions about federal defense investment and military bases.

The Politics of Military Base Closures: A Legislative History

Between 1976 and 1988, there were no domestic military base closures. Ironically, during these years the biggest base closure *advocate* was the Pentagon. As noted previously, the Defense Department has engaged in an ongoing assessment of the military needs of the nation and continually reviewed the military base structure in light of changing requirements. The round of military base closures after Vietnam had created strong congressional resentment. Members of Congress were upset because they had anticipated neither the broad extent of these closures nor their substantial cumulative economic and political impact.[41] Thus Congress came to view base closures negatively and in 1965 passed legislation requiring congressional approval of any Defense Department base closure program. The legislation was vetoed by President Lyndon Johnson but signaled the beginning of confrontation between the executive and legislative branches over military base closures. Members of Congress continued to seek more control over a process that they perceived to dramatically affect local interests in their constituencies. In 1976, Congress tried once again to clear legislation banning the executive branch from closing bases without congressional approval, but President Gerald Ford vetoed the measure. Congress, however, did succeed in enacting legislation (Public Law 94-431), commonly referred to as Section 2687, that required costly and time-consuming environmental-impact studies to be conducted on any bases slated for closure—a mechanism that effectively discouraged the Defense Department from closing bases.[42] Recommendations for military base closures still fell within the purview of the executive branch but were now subject to congressional notification (via environmental- and economic-impact reports) and thus to filibustering. This legislative maneuvering caused a ten-year stalemate on military base closures. It also kept open many obsolete or quasi-obsolete military bases, including the Presidio.

From 1976 to 1988, Defense Department officials attempted to close redundant or outdated installations but were frustrated by representatives and senators who sought to preserve local jobs and the extra federal aid those installations generated for local schools.[43] Perhaps Senator Phil Gramm, R-Texas, said it best: "There is something in the heart of every politician that loves a dam, a bridge or a military base."[44] In 1978, the Defense Department suggested eleven candidates for closure; Congress rejected all of them. In 1979, the department proposed nine candidates for closure; no closings were implemented.[45] In 1985, Reagan's defense secretary, Caspar Weinberger, suggested twenty-two installations be shut down in order to concentrate defense appropriations in areas defined as critical to national security.[46] Congress closed none. In 1987, Republicans offered to vote for the closure of twenty-eight bases; however, twenty-seven of these bases were in districts represented in the House by Democrats. The Democratic-led Congress closed none. By 1988, many politicians, including Representative Dick Armey, came to the conclusion that the process of closing military bases had become "too political for Congress to handle."[47]

The stalemate was politically effective for congressional members because it kept the economic and social disruptions from base closures from angering the voting constituency, but it meant that many military bases were kept open despite their obvious obsolescence. Representative Sam Gejdenson, D-Conn., explained, "When we look around this country, many military bases were put in place to protect settlers from the Indians . . . but the only things these bases are protecting these days are two Senators and a Congressman."[48] Fellow senator William Roth, R-Del., added, "we no longer need to guard the Pony Express routes to the Wild West. We no longer need to maintain moated forts to guard against assault by British sailing ships."[49] It became apparent that Congress had kept military bases open for political reasons—to maintain a flow of defense dollars into certain cities, states, and regions. This political posture was increasingly more difficult to maintain given the domestic economic pressures for budget reevaluation.

The actual process of closing military bases has had two *political* impediments: first, the parochial interest of each member of Congress who has a base in his or her district; second, a sense that members believe that bases closed in the past (notably the post-Vietnam closures) were closed for political reasons rather than for reasons of defense preparedness. Representative Carl Levin, D-Mich., commented: "The fear of the exercise of untrammeled executive power [to close bases] is what led or what continues to fuel the support for the protections against base closings."[50] The conflict between congressional duty to represent local interests and to set national policy is apparent with regard to base closures. In 1988, Representative Dick Armey, who recognized the parochial impulses of Congress, believed various factors had produced a window of opportunity in which to address this stalemate. He proposed a bill to create an independent twelve-member base closure commission to be appointed by both Congress and the executive branch.[51]

Armey's proposed legislation to create the base closure commission went before Congress in October 1988. The bill passed overwhelmingly and was signed into law shortly thereafter. In approving the Military Base Closure and Realignment Act of 1988, Congress acknowledged its virtual moratorium on base closures and the impasse caused by its own parochial impulses, and "it took itself out of the process."[52] In doing so, Congress proved it could subordinate constituency interests in preserving local payrolls to the broader national interest in trimming the deficit.[53]

Public Law 100-526 establishing the 1988 Base Closure and Realignment

Commission attempted to remove pork-barreling and partisan politics from military base closures by stipulating that Congress would be given only one choice: all or nothing, to accept or reject the commission's recommendations as a whole. *There could be no modification to the recommendations, only approval or total rejection.* Congress was thus confined to one yes-or-no vote on all the commission's recommendations, eliminating the ability of senators and representatives to bicker about each base. For any one member to succeed in saving a specific base, Congress would have to throw out the commission's entire list, an action that would halt the base-closing effort and thwart defense reduction plans. The political imperative of budget-deficit reduction prevented such action, and military base closures, once previously blocked by the political posturing of Congress, came about because Congress shifted responsibility for the decision-making process. This was, essentially, legislation by omission—removing the political onus of military base closures by passing the buck from Congress to the base closure commission. This did not mean, however, that "politics" had been removed from the decision making.

BRAC

The Base Closure and Realignment Commission (given the acronym *BRAC*) submitted its first report to the secretary of defense on December 29, 1988, recommending that 86 military facilities and properties be closed, 5 partially closed, and 54 realigned during a five-year period (1991–1995).[54] Of the 86 military bases, 16 were defined as major bases, and one of them was the Presidio Army Post. That the Presidio was one of the first bases recommended for closure reinforces the notion that the post had long been obsolete as an important military defense command center.

In its report, BRAC estimated the annual savings resulting from these closures and realignments to be $694 million a year and a twenty-year savings of $5.6 billion.[55] On January 5, 1989, the secretary of defense accepted all of the commission's recommendations and sent the list for congressional review. The legislation stipulated that unless Congress passed a resolution rejecting BRAC's recommendations, the plan would become law. Thus no congressional resolution of disapproval signified approval. The congressional review period expired without a joint resolution of disapproval (neither was there an overwhelming vote of approval). Ironically, the commission's list was signed into law by the architect of the biggest peacetime military buildup, President Ronald Reagan.[56]

Although the commission appeared to be supported by both political parties, this did not necessarily remove politics from the process of base closure. Congressional members whose constituencies faced a military base closure claimed the commission's list was politically driven by malevolence toward districts that had not been traditional supporters of the military. That claim was made by Senator Barbara Boxer, D-Calif., and Representative Nancy Pelosi, D-Calif., on behalf of the Presidio.[57] Some in Congress believed the commission had relied on misleading data with regard to actual defense savings and implied a Defense Department bias existed as well. The decision about which bases to close was politically neutral only if a senator's or representative's district was spared; it was politically charged in any district facing a base closure.

At the time, President Reagan and members of Congress believed that the 1989 base closures would realize significant defense savings and contribute to federal budget reductions on a scale that would postpone the need for further closures until the turn of the century. Senator Les Aspin, D-Wis., offered his interpretation of the base closure decision and assessed the political impact: "About two dozen members have been hit. More than a hundred are breathing a sigh of relief. They now have a vested interest in seeing the list go through, because that will give the bases in their areas at least a decade's breather."[58] The dramatic events in Eastern Europe and the Soviet Union, however, meant the original list was to be only the first of four rounds of post – Cold War base closings.

Round and Round We Go

In 1990, Secretary of Defense Cheney responded to continuing geopolitical events by proposing a more dramatic restructuring of the defense organization (which would include additional military base closures); he also requested and received legal authority to reactivate BRAC. Cheney's restructuring plan already included a list of some twenty-five military bases to be closed. Many congressional critics (who not coincidentally represented districts where bases were targeted for closure) claimed Cheney's new list unfairly targeted districts represented by Democrats; their response was to introduce new legislation to address these concerns (see below). More than 90 percent of the bases chosen by the secretary of defense were located in House Democratic districts (many of which had been perceived as hostile to the defense buildup of the Reagan years).[59] The commission responded by noting that "no effort was made to distribute the recommended closures geographically; the primary decision factor was cost-savings."[60] These congressional members teamed up with members affected in the 1988–1989 round to suggest changes in the way the commission would make decisions. Congress incorporated its members' criticisms by passing a modified version of the original act, the Defense Base Closure and Realignment Act of 1990 (Public Law 101-510).

This act initiated the "second round" of post–Cold War base closures. Similar to the 1988 legislation, it assigned an eight-member group to the commission and gave it more review power. The 1990 legislation also included three new provisions suggested by members who had been "hit" in the first-round closures: first, it replaced the "behind the doors" review with a more open review process consisting of public hearings; second, it required BRAC members to visit facilities proposed for closure; and third, it required that the General Accounting Office review the data and methods used by BRAC in making its base closure recommendations. The legislation also provided authority for BRAC to implement a total of three additional rounds of shut-down decisions, scheduled for 1991, 1993, and 1995.

These amendments indicate that although Congress had hoped to eliminate politics from base closure decisions, there seemed to be no way to prevent political elites from perceiving a political agenda attached to base closures. Still, few members of Congress were willing to publicly denounce the concept of defense reduction and reorganization; it had become a political imperative in the context of federal deficit reduction.

BRAC was mobilized again in early 1991 and by spring had selected another twenty-six major bases for closure and forty-eight for realignment. From the second round, the Defense Department projected savings of $2.3 billion in each fiscal year from 1992 to 1997.[61] This time, the projected savings was referred to as the "peace dividend." Once again, the congressional review period expired without a resolution of rejection, so President Bush signed all of the commission's recommendations into law in July 1991.

The Persian Gulf War of 1991 reemphasized the need for a strong, mobile, and prepared defense and raised questions about whether defense reductions would continue, given the increased volatility elsewhere in the world. However, defense budget reductions and military base closures did continue, even more dramatically. In January 1993, shortly before President-elect Bill Clinton was inaugurated, President Bush appointed the 1993 BRAC, pursuant to the 1991 legislation, to recommend the third round of base closings in five years. In March 1993, the commission released its recommendation that 28 additional major military bases be closed and 134 other facilities be cut back or consolidated. The Defense Department estimated that the third round of base closures would remove approximately 24,000 military jobs and 57,000 civilian jobs.[62]

Finally, in March 1995, the Base Closure and Realignment Commission informed President Clinton of its fourth round of decisions. Thirty-one major bases were targeted for closure, 84 for realignment (of which 25 were major bases) for a forecasted savings of approximately $1.8 billion per year and $18 billion over twenty years.[63] The Defense Department estimated that between 1985 and 1995 the number of service men and women would be reduced by one-third and defense spending would decline nearly 40 percent below the 1985 peak year in its 1996 budget.[64] Table 5.1 summarizes the four rounds of post–Cold War military base closures; Figure 5.4 details the cumulative effects of the four rounds.

In 1988, the U.S. government maintained 871 military installations and properties within the United States and another 375 overseas. It cost slightly under $17 billion each year to operate these bases. After the 1995 round of base closures, more than 548 (or 63 percent) of these 871 domestic installations had been closed or realigned for a six-year savings (or peace dividend) of $7.1 billion and a twenty-year savings of $56 billion.[65] And of the 548 bases affected, more than 102 are major bases that are slated for closure; this makes these post–Cold War base closures the most extensive in U.S. military history.

Table 5.1

Base Closure and Realignment Costs and Savings (in billions of FY 95 dollars)

Round	Closures/Realignments (number of installations)	Closure Costs	6-Year Net Savings	Total Savings*
1988	145	$2.2	$0.3	$6.8
1991	82	4.0	2.4	15.8
1993	175	6.9	0.4	15.7
1995	146	3.8	4.0	18.4
Totals	548	**$16.9**	**$7.1**	**$56.7**

Source: Adapted from U.S. Department of Defense, Base Closure and Realignment Commission, *Base Closure and Realignment Commission Report, 1995* (Washington, D.C.: GPO, March 1995).

*Net savings after closure costs, measured over 20 years and discounted to present value at 4.2 percent.

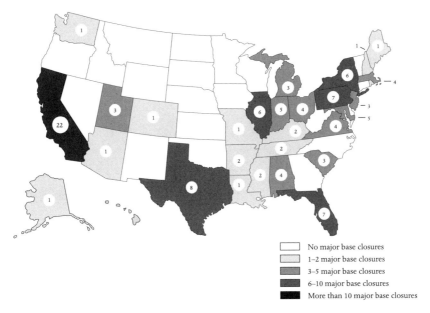

Figure 5.4
Cumulative Major Base Closures in Rounds 1, 2, 3, and 4 of post–Cold War Base Closures, 1988–1995. The word *major* signifies the loss of at least 300 military/civilian jobs. Totals: 548 facilities/installations affected, including 102 major base closures and 86 major base realignments, for a projected savings of $56.7 billion.
Source: Compiled by author from data in U.S. Department of Defense, Base Closure and Realignment Commission, Base Closure and Realignment Reports for 1988, 1991, 1993, and 1995.

SOME CONCLUSIONS

The mechanics and the politics of domestic military base closures are a result of a complex interaction among historical precedents, economic troubles, geopolitical changes, and legislative developments. This chapter has situated military base closures as the outcome of global and national political processes. The following chapter focuses on the consequences of a military base closure—the challenge of conversion. It examines the mechanics and politics of a military base conversion by documenting the process of planning for conversion and the end result—the "Grand Vision" of the Presidio's transformation from an Army post to a national park.

CHAPTER 6

Planning the Presidio's Conversion: From Post to Park

Tucked away on the back page of the *San Francisco Chronicle,* a small headline on October 10, 1988, announced, "50 Military Bases May Be Closed." In January 1989, when the Defense Base Realignment and Closure Commission's list of base closures was made public, the Presidio was listed as one of the twenty-six major military bases to be closed. In its report, BRAC estimated that closing the Presidio would save $50.2 million a year in Pentagon operating costs and offer a one-time savings of $313 million.[1] A few days later, on January 5, 1989, Defense Secretary Frank Carlucci approved the commission's recommendation, and the entire list then went before the Congress for a forty-five-day review period. When Congress did not reject the list, President Reagan signed it into law. This set in motion the Defense Department plan to vacate the Presidio by the fall of 1994. There would be a five-year transitional period in which to plan for its conversion.

The decision to close the Presidio set in motion the post's incorporation into the surrounding federal parkland, the GGNRA, as mandated by the 1972 legislation establishing the Golden Gate National Recreation Area. The Presidio's future landlord would be the National Park Service. Of all the military bases slated for closure in the 1988–1989 round, the Presidio was the only one mandated by federal law to become a park. The five-year transition period and the legal restrictions with regard to this now "surplus military space" initiated a complex conversion-planning process.

THE CITY REACTS TO BASE CLOSURE

In the San Francisco community, there were, broadly speaking, two reactions to the Defense Department plan to close the Presidio. One was elation, because the Presidio would finally be transformed into the long-anticipated park within the GGNRA. The other reaction was outrage, accompanied by denial, primarily based on economic factors. While many in the community were celebrating the inevitable transfer of the post to the GGNRA, local politicians rallied against the proposed closure. Ironically, Representative Nancy Pelosi and Senator Barbara Boxer, both liberals who consistently opposed wasteful military spending, led the rally. Boxer and Pelosi argued that costs associated with closing the Presidio could exceed those needed to keep it open. Specifically, they cited the potentially enormous costs of cleaning up the on-post environmental hazards as required by law for any federal land transfer. These cleanup costs, they argued, could make the Presidio more expensive to operate as a park than as a military facility.[2]

Boxer and Pelosi appealed to other congressional members to reevaluate closing the Presidio. Many local commentators, however, felt that Boxer and Pelosi were only posturing for the home crowd rather than putting energy into planning for the future. In the nine months following the announcement, Boxer and Pelosi lobbied to keep the Presidio exempt from the BRAC list, even though it appeared a lost cause. The Military Base Closure and Realignment Act was designed to prevent exactly this type of base-by-base negotiation.[3] In order to exempt the Presidio, Congress would have to reject the *entire* BRAC list. Such action was, at the time, politically infeasible. At the last moment, Boxer and Pelosi beat a tactical retreat when it was clear that there was no political support for their position.

Curiously, there were no grass-roots or local movements to save the Presidio (other than efforts by retired Army officers who had served at the Presidio).[4] Most civic energies had long since turned to the challenge of envisioning the Presidio's future. Several newspaper editorials and letters to the editor criticized Boxer and Pelosi for ignoring the immense potential for community benefits from a demilitarized Presidio, calling their efforts a "fight for headlines" but not results.[5]

Boxer and Pelosi's fight to save the Presidio came at the expense of lobbying efforts they could have made to secure federal funds to aid the National Park Service in the planning process. Park Service officials found themselves in a difficult position: on the one hand, they needed several million dollars of federal funds to move the planning process forward; on the other hand, they were wary of undercutting the Boxer and Pelosi strategy of contesting the base closure (aware that any plan for the Presidio would ultimately require the politicians' endorsement). Indeed, some felt that the unspoken message to the Park Service was to do the minimum necessary and not get out in front of the political fight to save the Presidio as an Army post. This made for a very frustrating year for the Park Service. Some members felt that an entire year was lost while politicians fought the closure.[6] When Boxer and Pelosi abandoned their staunch lobbying efforts for the Presidio's exemption, they joined local and regional efforts to increase the National Park Service budget in order to pay for the surveys and inventories of the post as well as the planning process.

A New Future for the Presidio

As a new park within the National Park Service, the Presidio would remain in the public domain as a uniquely open Army post. Although the Burton law stipulated that the Presidio be incorporated into the GGNRA, the law did not specify what type of park the post should become. The Park Service had no guidance as to what the Presidio should look like, what changes should be made, or how to incorporate the Presidio into the larger GGNRA.

Since the impetus for closing military bases around the country reflected concerns about the spiraling national debt and federal budget deficit, some public officials and citizen groups such as People for the Golden Gate National Recreation Area worried that the 1972 Burton law would be overturned to allow the Army to sell off the Presidio for private development, in order to generate money for the Defense Department. This was not a farfetched idea, as a recent struggle to save Alcatraz from a similar fate reminded many.[7] Several members of Congress threatened to sponsor legislation that would allow the Presidio to be sold to private business. Most San Franciscans had long ago recognized that the Presidio had little value as a military post, and there was little disagreement about its potential real estate value. The 3 square miles of spectacular ocean and bay vistas and the hundreds of potential view lots could generate millions of dollars in private development projects.[8] A residential development project at the Presidio would likely rival projects in adjacent prestigious neighborhoods. Locals wrote to the newspapers to express their worry that the outcome of private development could take nightmarish forms such as subdivisions of $2 million "monster houses" in a gated community, commercial development accompanied by Winnebagos cruising the parking lots, and kiosks for selling T-shirts.[9] They feared the resource would be parceled out and "degraded by a strange melange of opportunistic politicians, middlebrow bureaucrats, populist nuts, arts and crafts innocents and Fisherman's Wharf concessionaires with their spectacularly tacky souvenirs."[10] Some foresaw a stampede of kitsch.

But there were many who more optimistically envisioned a glorious future for the Presidio. They worried less about the possibility of private development and focused instead on the Presidio's mandate to become part of the GGNRA. The announcement of the post's closure renewed public interest in the Presidio's history and its future as part of the GGNRA. In the days and weeks following the closure announcement, a deluge of media attention featured articles on the historic Presidio and its future conversion. A few days after the announcement of the Presidio's impending closure, the *San Francisco Chronicle* ran a four-page article titled "The All-New Presidio: 1001 Ideas on What to Do with It Now." Another newspaper, the *San Francisco Independent,* encouraged residents to "Help Shape the New Presidio." The *Chronicle* next sponsored a Presidio Idea Contest. A month later, the newspaper had received more than five hundred ideas about the Presidio from local school children, Bay Area residents, architects, visitors, and Army veterans.

The Presidio's complex multiuse site received the attention of a variety of special interest groups. For example, the nonprofit San Francisco Friends of the Urban Forest submitted plans to showcase the Presidio's 400,000 trees as a model urban forest management program; a local boardsailing association lobbied to preserve Crissy Field's dunes and access to the bay; the Nude and Natural Society lobbied to retain the "clothing-optional" status of Baker Beach; and the Mycological Society of San Francisco implored the Park Service to overturn the Army's ban on mushroom picking in the Presidio.[11] Mikhail Gorbachev declared the Presidio an ideal site for the future headquarters of the U.S. chapter of the Gorbachev Foundation.[12]

In addition to the one-of-a-kind ideas, a majority of suggestions called for lots of open space, exactly what the long-term GGNRA scheme intended. No matter what the proposed project, the environment or nature was the unifying theme, as one response captured: "We like the idea of turning guns into butterflies and meadows."[13] "The Presidio inspires me to dream," said San Francisco State University president Robert Corrigan, "that within its 1400 marvelous acres, we could, without damage to park land, establish an education park to benefit all the citizens of our city."[14] And Kevin Starr wrote, "The Presidio is a prophetic place where the future is evoked and struggled for in ways at once symbolic and practical; it is arguably, the most impressive piece of relatively undeveloped urban real estate in America."[15] Long a special place to many San Franciscans, the Presidio inspired impassioned and poetic tributes. It became clear that in San Francisco, plans for the Presidio's conversion would not be greeted with indifference. Rather the input, suggestions, and participation of city residents would constitute an important element in the planning process.

The theme of maintaining the Presidio as a park quickly emerged as the dominant idea. A deluge of letters to the editor and editorials in the city's newspapers made clear that many San Franciscans wanted the Presidio to continue to provide open space and recreation. This would be not a radical transformation of space but rather the enhancement of a long-held tradition.

A few people were tempted to view the Presidio as offering solutions for some of the city's urban ills. There were several suggestions that the Presidio's facilities could provide affordable housing, shelters for the homeless, an AIDS hospice, and a recovery center for drug addicts. Indeed, the Presidio's complexity and its different land-use patterns lured some to see the post as a series of small spaces and opportunities rather than as one large space to be comprehensively planned. Although many residents were attracted to suggestions that the Presidio could be a panacea for the city's troubles, many more saw the future of the Presidio as a grand opportunity. Aside from worries about development or piecemeal planning, there was a remarkable consensus that the Presidio should continue to provide open space, recreation, and spiritual renewal. There was also general agreement that the Presidio should become something more than just another park area within the GGNRA. Bay Area resident William Penn Mott, former director of the National Park Service and an active participant in the Presidio planning process, urged the community to aim high:

> The Presidio's location and facilities are such that *we have to look at it from a bigger point of view than just a national park*. It can and should be a global resource. What should the big picture for the Presidio be? Where is the vision that will stir our blood, that will be commensurate with the unique quality, beauty and inspiration of this world-class site? What magic wand can be waved over this special place that will transform it into a noble achievement destined to live forever [my emphasis]?[16]

So began the quest for a Grand Vision worthy of the Presidio.

Mobilizing the Community for the Planning Process

Early in the process, it became apparent that the Park Service had neither the money nor the expertise to create a national park out of a military base the size of a small city. At that point, it enlisted the help of the Golden Gate National Park Association (GGNPA), a private nonprofit organization and park partner that funds volunteer and educational programs for the GGNRA.[17] The Park Service asked GGNPA to help with ideas about how the planning process should be developed, how to incorporate the public voice in planning, and what kinds of expertise should be brought to the planning. Out of these early discussions, GGNPA developed the concept of a new nonprofit organization: the Presidio Council.[18] Three of the driving forces behind the creation of the Presidio Council—GGNPA chair Toby Rosenblatt, GGNPA executive director Greg Moore, and James (Jim) Harvey[19]—believed that although the Park Service could obtain the input of experts through informal methods, volunteers would not be fully engaged without a formal organization. They explained:

> If we want exceptional volunteers of regional and national stature, their involvement must have substance and symbolism.... the work of the Task Force will help provide credibility to the eventual recommendations of the plan. It will verify that we have sought the best minds and guided the plan to visionary concepts.

GGNPA foresaw the Presidio Council as an advisory group that would bring national perspective to the planning process and provide guidance on management and finance.[20] It would be composed of national leaders in urban planning and design, environment, education, finance, public health, facility management, historic preservation, cultural affairs, engineering, social science, medicine, travel and tourism, architecture, and landscape architecture.

In April 1991, GGNPA selected a list of experts and invited them to join the Presidio Council (see Appendix B for a list of the members). Council members would assist the Park Service as pro bono consultants in four broad areas of planning and analysis. They would help refine planning concepts; they would suggest a framework for evaluating various "visions" and "alternative concepts" for the Presidio's future; they would provide expertise in financial feasibility and urban planning; and they would assess the financial feasibility and management structure for implementing Presidio programs and plans.[21]

The council would also provide access to other individuals or institutions that could help provide support to the planning effort. To this end, the Presidio Council and GGNPA worked together to raise more than $956,000 from private and corporate sources and leveraged over $1 million in pro bono support to conduct economic and real estate analyses.[22] The privately raised money was divided in three directions. First, a portion of it allowed GGNPA to hire a staff experienced in planning, communications, and government relations to aid the Presidio Council. Second, some money went for communications—a newsletter, summaries of the plan, press communications, as well as the glossy brochures sent out to recruit potential tenants. Third, some money went to fund consulting studies that GGNPA completed on behalf of the planning effort. Indeed, the Park Service's lack of urban planning expertise and its chronic shortage of funds made the Presidio Council an invaluable participant in the decision-making process. There were, however, occasional critics of the Presidio Council. The independent *Bay Guardian* ran several articles criticizing not only the Park Service but also the Presidio Council, referring to the group as "a few powerful executives with special-influence peddling."[23] Despite these minor criticisms, the Park Service acknowledged the great debt owed to the Presidio Council and GGNPA for their assistance in the planning process.

In addition to the Presidio Council and GGNPA, the GGNPA Advisory Commission, an eighteen-member commission appointed by the secretary of the interior, also assisted the Park Service in the planning process.[24] It would review and finalize the planning guidelines. These were the most visible and influential nonprofit, non–Park Service organizations involved with the Presidio planning process. In addition to these formal organizations, many project consultants were hired by the Park Service to conduct land and building surveys, transportation analysis, and financial and investment studies.

Mobilizing the Park Service for the Planning Process

Although the GGNRA had a complete staff of planners and resource managers, GGNRA officials understood that the demands of putting together a master plan for the Presidio would require a staff devoted exclusively to this task. They could not afford one, however. Instead, GGNRA officials and other Park Service officials assembled a locally based Presidio Planning and Transition Team, staffed by employees specializing in park planning, historic preservation, landscape architecture, law, finance, and community development. The Park Service also provided funds for a San Francisco city planner to serve on the team to ensure coordination with all relevant city and county departments and officials.[25] The Presidio planning team would consist of a core team that would be stationed at the Presidio throughout the planning effort and an extended team that would include additional people from other components of the Park Service who would meet periodically to review the planning effort. The core team of seven people was led by the planning team captain Don Neubacher. The extended team had thirty-five members.

In assembling the Presidio Planning and Transition Team, Park Service officials drew on the expertise of the entire National Park Service staff, gathering specialists from the local GGNRA and importing others from elsewhere, including the Denver Service Center—the planning, design, and construction division of the National Park Service (see Appendix C). Brian O'Neill, GGNRA superintendent, noted that assembling a core team to be stationed at the Presidio throughout the planning process was an unusual measure for the Park Service but necessary because of the need to establish a long-term rapport with the city of San Francisco.[26]

Both the Park Service and the GGNRA Advisory Commission were committed to a highly participatory planning process. They scheduled regular public meetings and workshops and provided Park Service documents and guidelines to the public at no cost. Two publications, *Presidio Update* (a newsletter produced by GGNPA) and *Reveille* (a newsletter of the National Park Service planning team) kept the public informed of the planning process and notified the public of planning events, hearings, brown-bag discussions, and workshops.

Public participation is important in creating an effective plan, for although the planning process includes the use of much technical knowledge, good planning is also about public relations and communications. Community participation can result in a well-supported plan, so public meetings, workshops, and hearings are important aspects of planning. The Park Service had little trouble mobilizing public interest and participation, in part because of the special relationship of the Presidio and city residents. Early in the planning process, more than four hundred people showed up at the Marina Middle School public hearing, bringing with them their ideas and their concerns.[27] At public hearings various community leaders and citizens suggested ideas or discussed concerns. Neighborhood associations, local community groups, and city officials attended these meetings to discuss issues of public safety, housing, and the impact of the Presidio-as-park on nearby residential and business areas. Still others voiced opinions about the Presidio's future use. Much of this feedback was incorporated into the eventual plan as well as the planning timeline.[28]

Significant participation by local citizen activists and nonprofit organizations (such as the Presidio Council, GGNPA, and various neighborhood associations) presented new challenges to traditional Park Service planning. Indeed, the complexity of planning and coordinating among the Department of the Interior, the Park Service, the Army, and local organizations and citizen groups presented a unique challenge to Park Service planners. Along with the well-coordinated and often influential input of non–Park Service groups and individuals, the Park Service also introduced extensive public workshops, hearings, and mailings. "The Park Service has realized that this is an entirely different kind of thing than they have handled in the past, so they have been very open and have encouraged input from outsiders," confirmed James Harvey, San Francisco resident, CEO of TransAmerica, and member of the Presidio Council.[29] Indeed, many felt that the Park Service did everything it could to open up the planning process.[30]

The Planning Process

Park Service planners had never dealt with such a complex multiuse site. Said one park official, "we're experienced in wilderness areas and historic sites, but in the Presidio we're taking over an entire city."[31] Indeed, they were. As Brian O'Neill remarked:

> the Park Service has never dealt with a built environment of this magnitude and a location like this. The whole idea of how to reuse buildings for functions that are much more noble, broader in scope and more worldly is a new challenge. The normal ideas that would come out of a planning process in the Park Service are more in line with what to do with a dozen buildings, not 800 of them totaling more than 6.3 million square feet.[32]

In 1989, Interior Secretary Manuel Lujan toured the Presidio and confirmed "the Presidio is a different kind of problem for the Park Service to face. It is not like the Grand Canyon, a natural resource that identifies itself. Here we have 800 buildings to fit into a national park."[33] As the Presidio's new landlord, the Park Service was challenged with an ambitious task: not only was the agency required to develop park program themes, but it would also have to secure tenants for the hundreds of buildings and make millions of dollars in repairs required to meet state codes.

National recreation areas are operated under the legal and administrative mandates of the National Park Service. The National Environmental Policy Act of 1969 mandated environmental impact statements for all development projects. In addition, each park unit was required to create a general management plan as a way of organizing management strategies and determining the best ways to develop Park Service property. In 1980, the National Park Service had approved a general management plan and environmental analysis for the Golden Gate National Recreation Area.[34] The master plan for the Presidio would actually be an amendment to the more comprehensive general management plan for the GGNRA.

As required by law, the master plan had to provide guidelines for the use, management, and development of the new Presidio park over ten to fifteen years. The Presidio planning team considered the master plan as a general guideline rather than as an exact blueprint. Central to a Park Service master plan is the establishment of program themes. Program themes would determine appropriate Presidio programs and would aid the Park Service in selecting Presidio tenants. The Presidio planning team, along with the GGNRA Advisory Commission, envisioned a master plan that would contain the following:

- An outline of the Park program themes
- A proposal for a governance structure for administering park programs
- Identification of additional federal legislation needed
- A phased action plan for managing the conversion[35]

Because of the many Presidio facilities, the Presidio planning team concluded that private enterprises, in some guise, would be necessary to meet the needs of the Presidio. After all, the Army spent $45–60 million each year to manage the Presidio and pay for its military police patrols, plumbers, gardeners, and others.[36] In comparison, the annual operating budget for the entire GGNRA, now more than 72,000 acres, was only $10.5 million, less than half of what the federal government allocated to the Army for a space of 1,500 acres.[37] The Park Service concluded that several commercial ventures were needed to meet the maintenance needs of the Presidio.[38]

The Visioning Phase, 1989–1991

Following the closure announcement in 1989, the Park Service began what it described as the "Visioning Phase of planning" (Figure 6.1 summarizes the timeline of the Presidio planning process). The purpose of the visioning phase was to capture a vision that brought forth the big picture for the Presidio, as Park Service Director William Penn Mott, Jr., had earlier challenged. The first task of the Presidio planning team was to establish planning guidelines. To help generate this first important document, the Park Service scheduled a series of workshops and public events at which the Presidio planning team could gather ideas and listen to public suggestions and concerns. This is not to say that the process was without conflict: the planning team faced the challenge of integrating (and even rejecting) many of the diverse visions for the Presidio's future. However, the planning guidelines were broad enough to allow for many envisioned futures. In November 1989, San Francisco State University and the Park Service sponsored a two-day conference ("Think Big") for city planners, park planners, local architects, and landscapers to discuss preliminary concepts and formulate preliminary Presidio planning guidelines. In May 1990, the GGNRA Advisory Commission and the National Park Service adopted the Presidio Planning Guidelines, the first document in the planning process (see Table 6.1).

Table 6.1
The Presidio Planning Guidelines

1. Those components of the historic landscape that assure and enhance its historic integrity will be preserved.
2. Historic structures will be put to use while preserving their basic exterior and/or interior historic features.
3. The character and scale of the Presidio's open space will be preserved and wherever feasible enhanced.
4. National features of the post will be protected and wherever feasible restored.
5. The Presidio will continue to be open to the public and public uses will be encouraged.
6. New uses in the Presidio should support or be compatible with National Park purposes as well as the significance and character of the site.
7. Appropriate uses that support the operating costs of the Presidio may be sought.
8. The Presidio planning process will take a long-term view.
9. The planning process will include ample opportunity for public input and coordination with local and state governments.
10. The Presidio's infrastructure will be modernized; hazardous wastes will be removed; and air and water quality, and other environmental values of the post will be restored and enhanced.

Source: National Park Service, *Presidio Visions Kit* (San Francisco: GGNPA and GGNRA, Winter 1991).

Figure 6.1
The Presidio Planning Process: A Timeline of Key Events and Decisions

1989–1990 The Visioning Phase	1991–1992 The Production Phase	1993–1994 The Plan Completed
January 5, 1989 Secretary of Defense announces Presidio closure. Park Service begins "visioning phase." San Francisco media blitz community for Presidio ideas. *April 1989* Park Service sets public hearings and workshops. *November 1989* Park Service Conference "Think Big." *May 1990* Presidio Planning Guidelines developed, reviewed by public during workshops. Park Service solicits ideas from public, national advisers. Park Service holds Presidio "Visions" workshops and distributes visions kits to obtain public suggestions/input.	Park Service "production phase" begins. Presidio Concepts Workbook published. Park Service considers preliminary themes, uses. Public workshops elicit comments on ideas. *November 1991* Park Service narrows Presidio use concepts. *April 1992* Park Service distributes Call for Interest to gather preliminary statements of interest in Presidio tenancy and gets 400 responses. First "market test" of Presidio vision.	*March 1993* Draft plan developed; environmental-impact statement and alternative options developed. Army turns over Presidio Forest, Lobos Creek Valley, and Coastal Bluffs to Park Service management. Draft plan amendment released for public review. Public hearings held. Revisions to draft. *September 1993* Army hands over cavalry stables, Crissy Field, and Army Museum to Park Service. *October 1993* Revised draft plan amendment published and several supplementary studies also published. *March 1994* Park Service takes over Presidio housing. *October 1, 1994* Transfer ceremony. All remaining areas of Presidio transferred to Park Service.

With the approval of the Presidio Planning Guidelines the Park Service felt it had "established the common ground upon which we can build our dreams for the Presidio."[39] The Presidio planning team was ready to begin gathering suggestions and comments about the transformation of the Presidio. To assist in the visioning phase, the Park Service/GGNPA distributed a *Presidio Visions Kit* to the public. The *Visions Kit* invited Bay Area residents to share their ideas for the Presidio and to "describe what you would like the Presidio to become." It introduced the various features of the Presidio, outlined the ten planning guidelines, and provided a worksheet for people to submit their ideas.

In conjunction with distribution of the *Visions Kit,* the Presidio planning team held workshops to encourage public participation. More than five hundred people participated in the eight "Visions" workshops held throughout the Bay Area in January and February 1991.[40] The Presidio planning team organized these workshops in the style of a town meeting. First, participants worked together in small groups to map out ideas for the various areas of the Presidio; then each workshop concluded with the small groups sharing their ideas with all participants; finally, the ideas generated at the workshops were combined to create one large, inclusive planning map.[41] The *Visions Kit* distributed earlier provided a basis for discussion and design.

In June 1991, the Park Service held a two-day trade show called the "Presidio Forum" to publicize Presidio ideas. Hundreds of completed vision kits, the planning maps designed during the "Visions" workshops, as well as informal suggestions and ideas were displayed. These displays and summaries were then photographed and made available to the public in a booklet entitled *The Presidio Forum Displays.* In December 1991, the Park Service released *The Presidio Concepts Workbook,* a booklet containing sample plans that had been generated during the visioning phase and the public workshops. The workbook marked the end of the visioning phase of planning and the beginning of the production phase—the creation of the master plan for the Presidio.

The Production Phase, 1991–1993

The Presidio Planning and Transition Team narrowed the goals for the conversion of the Presidio to four broad concepts: (1) stewardship and sustainability; (2) cross-cultural and international cooperation; (3) community service and restoration; and (4) health and scientific discovery (see Table 6.2).

Table 6.2
Presidio Park Program Use Concepts

Stewardship and Sustainability
Objective: Promote and advance research, education, training and demonstration and policy formulation on major environmental issues of worldwide importance using Park Partners.

Cross-Cultural and International Cooperation
Objective: Build on the historical role of the Golden Gate as a crossroads of international exploration, cooperation and exchange.

Community Service and Restoration
Objective: Convey the value of public service by promoting responsibility, leadership, stewardship and community participation in improving the places where we live, work and play.

Health and Scientific Discovery
Objective: Promote life and earth science research, emphasizing systems and methods to improve human health and the quality of the environment for future generations.

Source: National Park Service, *Creating a Park for the 21st Century: From Military Post to National Park. Draft General Management Plan Amendment, Presidio of San Francisco,* NPS D-148 (San Francisco: GGNRA, October 1993: 26–27).

These use concepts would coalesce into the broader "Grand Vision" that would help to define management and use of the post.

Two of the land-use concepts (community service and restoration; cross-cultural and international cooperation) were ones with which the Park Service had considerable administrative and management experience at national parks and historic sites. The other two (health and scientific discovery; stewardship and sustainability) presented new challenges to the Park Service, because they were unique in scale and scope for a national recreation area, or indeed for any of the lands within the national park system. Importantly, these four use concepts identified the future Presidio as more than a single-purpose space; rather, the broad concepts astutely took advantage of the complexity of infrastructure as well as existing land-use patterns. The program themes appeared to be well received, although several neighborhood organizations and the media complained that the preliminary plan was being rushed through public review.[42]

In April 1992, the Park Service issued the *Call for Interest* proposal to solicit local, nationwide, and even international responses from nonprofit organizations, academic institutions, government agencies, and for-profit enterprises wishing to occupy space at the Presidio. The document described the features of the Presidio, listed the specific resources available to prospective tenants, and identified the Presidio Planning Guidelines. More than five thousand documents were sent out; the Park Service received more than four hundred responses.[43] The *Call for Interest* was the first attempt by the Park Service to gauge the interest level of prospective Presidio tenants.

For three days during June 1992, the Presidio planning team organized a design workshop to generate ideas from community landscape architects, planners, and architects for the Presidio. Many of the participants were either Presidio Council members or Presidio pro bono "Park Partners." Although the Park Service and GGNPA had sought extensive public involvement at the workshops and public hearings, the planning team and citizen activists worried that many groups were underrepresented in the planning process. The Presidio Council, in conjunction with the Park Service, launched the "Community Consultation Initiative," a forum for community concerns. More than twenty of the Bay Area's leaders of urban neighborhood associations and community organizations attended the meeting and gave their input into the planning process.[44]

During the rest of 1992, the Presidio planning team continued to organize or attend various public hearings and workshops in order to elicit and encourage public input. At the same time, the team was drafting the master plan document.

THE PRESIDIO'S "GRAND VISION": AT LAST, A PLAN

On October 19, 1993, after three years of planning, the Park Service unveiled its blueprint for the conversion of the Presidio. The draft master plan was the outcome of several years of planning, a process that opened numerous avenues for the public exchange of ideas and for expert consulting and analysis.[45] The master plan, or "Grand Vision," consisted of the primary plan amendment and five supplementary documents. Several of these supplementary documents, such as the environmental-impact statement and the transportation planning and analysis technical report, were required by federal law. Several of these analyses and supplementary documents were prepared by consultants hired by the Park Service and provided the Park Service with a detailed assessment of the post's facilities and structures—an "inventory" for the planning team to work with.[46]

The plan called for the Presidio to be "a great urban national park; a center for research and learning; a model of sustainability."[47] The proposal also called for "Park Partners" to work with the Park Service to create innovative park programs that reflected the four general use concepts.

As a way of dealing with the complexity of the Presidio and its multiuse character, the Presidio planning team divided the Presidio into thirteen geographic or planning areas (see Figure 6.2). Park planners had identified these thirteen land-use areas based on a combination of topography, views, established land-use patterns, and public input (from the *Visions Kit*). In introducing the Grand Vision, it is useful to sketch briefly the highlights of the plan for each of these thirteen Presidio planning areas.

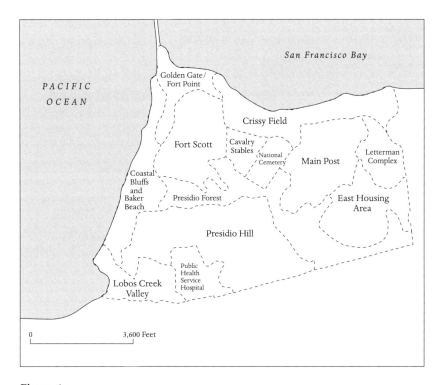

Figure 6.2
The Thirteen Planning Areas of the Presidio. *Source: Map by author, adapted from National Park Service,* Creating a Park for the 21st Century: From Military Post to National Park. Draft General Management Plan Amendment, Presidio of San Francisco, *NPS D-148 (San Francisco: GGNRA, October 1993).*

1. The Main Post The site of the original Spanish presidio, the Main Post has historically functioned as the administrative center of the Presidio. Several clusters of historic buildings surround the two parade grounds that define the central space. According to the Grand Vision, the Main Post will remain the focus of activities and the center of the Presidio community. Although all four program concepts are equally important to the overall plan, the jewel of the Grand Vision rests on the theme of stewardship and sustainability. Thus the highlight of the Grand Vision is the transformation of the Main Post into a global environmental center at which national and international institutions will sponsor workshops, educational activities, and joint studies on issues related to major environmental and societal challenges.[48] Along with plans to situate the global environmental center in the buildings surrounding the parade grounds at the Main Post, many of the existing facilities will provide community services to Park Service and Park Partner employees (these include a child care center, post chapel, theater, post office, and commissary).

In addition, the GGNRA Orientation and Visitor Center will be relocated from nearby Fort Mason to the Main Post, a sign that the center of the GGNRA will be at the Presidio. To maintain the Main Post's centrality, the Presidio planning team proposed transforming the area into a pedestrian district.[49] In conjunction with this plan, one of the more interesting and innovative ideas for the Main Post involves the removal of the paved parking areas on the main parade ground. More than 640 of the 1,100 parking spaces will be removed and the area restored to grassy lawn.

2. Fort Scott Often described as the most architecturally coherent set of buildings at the Presidio, Fort Scott is a cluster of approximately 159 Mission Revival buildings and includes barracks, offices, a gymnasium, a ballfield, a parade ground, and three separate housing clusters. The coherence and strong sense of unity of Fort Scott has resulted in an area with a distinct style and campus-like atmosphere. The Presidio planning team took advantage of this architectural and spatial unity by proposing that Fort Scott be transformed into a conference, training, and research activity center that will focus on environmental and other global concerns. Programs at Fort Scott will highlight environmental education and training and will address applied rather than basic research in sustainable development, human ecology, natural resource preservation and management, and cultural resource preservation. The Mission Revival barracks, which are grouped in a horseshoe-shaped parade ground, will serve as the conference and lodging center; nearby warehouses and offices will provide space for classrooms, workshops, laboratories, and training rooms.

3. The Letterman Complex Located on the far east side of the Presidio, the Letterman Complex is dominated by the Letterman Army Medical Center and the Letterman Army Institute of Research (known as "LAIR"). The Letterman Complex is a specialized 60-acre biomedical campus consisting of laboratories and clinical space. It is anchored by the ten-story hospital constructed in the 1950s. This is perhaps the most modern and urban area in the Presidio; it presented one of the most difficult challenges to the Park Service planners as they tried to envision a use compatible with the Presidio vision. In fact, this planning area was so unusual that the Park Service dealt with it as a separate planning issue.

In July 1993, the Park Service hosted a two-day conference to which they invited a distinguished group of participants with expertise in public health, developmental biology, biomedical research, and the environment. A variety of expert volunteers formed the Letterman Design Group and assisted the Park Service in envisioning how the unusual and specific site of Letterman should be developed and managed.[50] Several critics suggested removing Letterman from the Presidio—selling it to the city or state, for example. But during the conference, it became clear to many, including the financial consultants and the design group participants, that the successful incorporation of the complex was necessary to the overall plan for the Presidio, because the high-tech, state-of-the-art laboratories at LAIR could command top-market leases.[51] Many people involved in the planning process worked to ensure Letterman's inclusion because the revenues from an anchor tenant at Letterman could generate the necessary revenue to subsidize the enormous costs of rehabilitating other Presidio buildings and could also subsidize the below-market leases that would be given to nonprofit organizations, including the crown jewel, the global environmental and education center. The difficulty of transforming the Letterman Complex into a new type of space that would attract a top-paying anchor tenant while ensuring compatibility with park program themes led to the suggestion that the program use concepts be revised to reflect "geographic reality."[52] The "health and scientific discovery" use concept emerged as a solution to the challenge. Letterman was the only one of the thirteen Presidio areas with the program theme "health and scientific discovery." This is an important aspect of the plan, because it reflects a broader reckoning with the economic realities of establishing the Presidio as a great urban national park, as much as a desire to retain Letterman within the Presidio. Thus the economic rationale dictated by the opportunities represented by LAIR led to a partial revision of the Grand Vision.

The Grand Vision calls for the ten-story hospital to be removed to enhance open space; LAIR will be dedicated to research and education focusing on issues of human health and on health concerns related to environmental hazards. One of the first responses to the *Call for Interest* came from the University of California–San Francisco, which expressed interest in being the single tenant of LAIR. In fact, during the entire production phase, Park Service officials, GGNPA, and the Presidio Council not only knew of the university's interest in LAIR but unofficially and informally negotiated the space's use by the university.[53]

4. Crissy Field Originally a long stretch of wetlands, present-day Crissy Field is a filled-in area located where the Presidio meets San Francisco Bay. It is currently the most public (and utilized) of the Presidio open spaces. Although a large section of Crissy Field is open space, one hundred buildings (some historic, others industrial) are concentrated at the west end. The Golden Gate Promenade, which extends along the length of the Crissy Field shoreline and ends at the Golden Gate Bridge Plaza, is the city's most popular running, biking, and walking path. The Grand Vision conceives of Crissy Field as the "front yard" for the Presidio.[54] The historic airfield structures will be rehabilitated and a museum of aviation technology established. Many of the nonhistoric structures will be removed to expand open space and to improve access to the waterfront. Park planners will design special access areas to the offshore waters for boardsailers. The historic (circa 1920) grass landing field will be restored and used as open space for large public events. The most innovative aspect of the plan for this area is a proposal for reclamation of the original Crissy Field wetlands. Reclamation efforts will entail the removal of nonnative and nonhistoric vegetation and the elimination of nonessential paved areas and parking lots. Portions of the wetlands along the bay shore will be restored. The reclamation or restoration of Crissy Field is an ambitious and potentially expensive project. Local philanthropic donors have pledged to help the Park Service with the ambitious plan by privately raising funds.

5. Golden Gate Bridge/Fort Point This planning area consists of coastal defense fortifications and the Golden Gate Bridge Plaza, which is the primary destination for viewing the entrance to the bay and the bridge. According to the San Francisco Tourist Board, the plaza has been the single most "visited" area of San Francisco, receiving some 20 million visitors by car, bus, tour bus, and bicycle each year. The plan for this area is to enhance scenic vistas and the historic setting. Nonhistoric buildings housing bridge administrative functions will be relocated so that there are more open areas surrounding the plaza. Several of the poorly defined hiking trails at the plaza that connect to other Presidio trails will be improved, pedestrian access will be improved, and exhibits at the Golden Gate Bridge Plaza will be built to provide a Presidio orientation for the millions of visitors as well as to the entire GGNRA.

6. Cavalry Stables/Pet Cemetery This small planning area contains only sixteen buildings and is nestled between Crissy Field and Fort Scott. The plan scripts only slight modifications to this area. Park Police will stable their horses in two of the cavalry stables; another stable will be rehabilitated and used as a museum that will document the history of the Army's mounted patrol service in California's early national parks. The pet cemetery will continue to be used by the Presidio community.

7. National Cemetery The San Francisco National Cemetery, the resting place of soldiers and their families, as well as the first national cemetery established on the West Coast, is administered by the Veterans Administration. The Park Service will maintain the cemetery; the Veterans Administration will continue to manage cemetery operations.

8. East Housing Area This area contains 135 historic and nonhistoric buildings, predominately housing for officers and enlisted personnel. Many of these houses are charming, built in either Dutch Colonial Revival or Mission Revival style. The East Housing Area will be maintained as a residential area for employees of the Park Service and Park Partners. The Park Service hopes this on-post housing will promote "sustainable living" through the reduction of commuting traffic and automobile pollution.[55]

9. Public Health Service Hospital This 37-acre complex is located on the southern boundary of the Presidio. The Presidio planning team proposed that it be transformed into an educational and conference center dedicated to either youth job training or the youth conservation corps. The hospital will be refitted to serve as a dormitory-style residence for the short-term program participants.

10. Presidio Hill This planning area is dominated by open space and includes Mountain Lake, the campsite of the first Anza expedition into the Bay Area. The 160-acre Presidio Golf Course and several small housing areas also lie within the planning area. The plan proposes to retain the recreational function of this area for bikers, golfers, and other visitors. Several of the nonhistoric housing communities will be removed and the site replanted with native plants. The Presidio Golf Course will become a public course with no private membership (San Francisco currently has six private golf courses but only two public ones).

11. *The Presidio Forest* The result of the tree-planting project initiated in the 1880s, the Presidio Forest is now a mature hundred-year-old forest. A Park Service analysis noted that "because of its historical and recreational significance, the Presidio forest will be revitalized and managed as part of the cultural landscape." [56] The Presidio Forest will be maintained as a woodland retreat in the midst of its urban setting. However, because many of the eucalyptus trees have reached the end of their life span, the forest will require extensive management and replanting to survive. Although many of the trees and plants are non-native, planners decided to retain the original 1880 design and will continue to replant eucalyptus trees. Several of the grassland areas support native wildflowers and rare plants, including the state-listed Presidio clarkia. The Park Service will clear these grassland areas of non-native plants (this conforms to regulations under the 1973 Endangered Species Act).

12. *Lobos Creek Valley* The creek has historically provided water for the Presidio through a series of underground springs and a subterranean aquifer. The creek flows along the southern edge of the Presidio and meets the ocean at Baker Beach. For many years, however, the creek has been diverted into a water treatment plant, and drainage pipes have disturbed the naturally flowing stream. Similar in concept to the restoration of Crissy Field, the Park Service plan calls for restoration of Lobos Creek to a flowing stream by rebuilding the stream bed and restoring the riparian corridor. When restored, planners believe, the creek will provide the Presidio's water needs by contributing between 1.8 and 2 million gallons of water a day.

13. *The Coastal Bluffs and Baker Beach* This is the least developed planning area of the Presidio. It provides habitat for several rare and endangered native plant species. The Park Service plans to preserve this area as "the wildest part of the Presidio landscape." [57]

The Presidio planning team paid considerable attention to the economics of the plan in an attempt to balance visionary ideas with economic realities. The cost of implementing the Grand Vision was estimated to be between $590 and $666 million (see Table 6.3).

Table 6.3
Estimated Costs of the Presidio's Grand Vision (in millions of 1993 dollars)

Cost Estimate Summary	
Construction and Development	
Main Post	$172
Fort Point/Golden Gate Overlook	11
Fort Scott	110
Letterman Complex	121
Cavalry Stables	16
Public Health Service Hospital	70
East Housing	15
Crissy Field	45
Presidio Hill	17
Cemetery	2
Presidio Forest	5
Lobos Creek	4
Coastal Bluffs	2
Subtotal	$590
Asbestos abatement	30
Studies/environmental-impact analysis	2
Infrastructure upgrades (parking lots, roadways, grounds)	44
Totals*	$666
Envisioned Revenue	
Building and facility rental from tenants	345
Federal government appropriations**	—
Private philanthropy***	245
Total Revenues	590

Source: National Park Service, *Creating a Park for the 21st Century: From Military Post to National Park. Draft General Management Plan Amendment, Presidio of San Francisco,* NPS D-148 (San Francisco: GGNRA, 1993: 112–113 and Appendix F, 138).

*Totals do not include hazardous waste removal (Army responsibility) or the Doyle Drive (Highway 101) upgrade and reconstruction (California State Department of Transportation responsibility)

**Estimated $24 to $28 million per year until 2010 to be requested from Congress.

***Estimated to be $30 to $40 million over fifteen years.

THE CONTINGENCY PLANS

Although the Grand Vision was a document, it was not a static scenario for the future. As the Presidio Planning and Transition Team intended, the plan was a general blueprint representing not only the end result of the planning process but also the beginning of geographical transformation. And this is an important point to keep in mind when considering the Presidio's master plan.

An effective way to gauge the degree to which the plan was truly a master plan in the traditional planning sense is to consider what possibilities the Grand Vision included or excluded. This is an easy task because the Presidio planning team formulated three alternative plans for the management and use of the post. These alternatives were to provide a series of contingency plans if the proposed Grand Vision failed to muster the needed support (financial and otherwise). The planning team hoped that by offering several alternatives it could ensure that at least one of the Park Service plans would be approved (thereby giving the Park Service control over the future of the Presidio) and convince both the community and Congress that the planning process had been thoughtful and thorough. The alternatives differ primarily in overall management, the extent of resource preservation and enhancement, and the scope and diversity of visitor programs.

Alternative B proposed that the Park Service implement only those actions necessary to meet the legislative directive to protect the Presidio's natural and cultural resources, as stipulated by the 1972 Burton law. The Park Service would carry out only limited cultural landscape restoration. No buildings would be removed. Crissy Field and Lobos Creek would not be restored. Most Park Service programs, services, and resources would be concentrated in open space areas. Any buildings leased at the Presidio would become the responsibility of the General Services Administration, not the Park Service. The Public Health Service Hospital would be excluded from the Presidio.

If the Grand Vision reflected planning for the transformation of the Presidio, alternative B represented planning for a transfer. This is a subtle but important difference. The word *transfer* implies that an object or place changes ownership, location, or responsibility, not necessarily form, function, or purpose. The word *transformation,* in contrast, implies a change in form, shape, substance, character, and meaning. If the Grand Vision proposed the Presidio's transformation, alternative B represented the fewest, most minimal changes to the Presidio possible, in keeping with legislative requirements.

Alternative C focused on the expansion of Presidio open space. This alternative would provide visitors with experiences like those they would expect in a traditional national park.[58] Under alternative C, the Park Service would increase open space, remove nonhistoric buildings, restore historic buildings, and protect native plant habitat. As proposed under the Grand Vision, the Park Service would coordinate the restoration of Crissy Field and Lobos Creek. There would be a variety of interpretive and educational programs at the Presidio, and the Park Service would have sole responsibility for administration, resource management, and interpretation, maintenance, and public safety. The Public Health Service Hospital and the Letterman Complex would be excluded from the Presidio boundary (special legislation would be needed to allow the sale of these two areas). This would limit the subsidies available for nonprofit Park Partners. Alternative C would establish Park Partners to operate under historic leases, cooperative agreements, and special use permits to establish a wide range of programming, but there would be no global environmental and education center. Alternative C would provide significantly more visitor opportunities than alternative B, though fewer than the Grand Vision.

Under alternative D, the Presidio would be partially reused by the military. Most of the Presidio would remain under National Park Service management, but the Department of Defense would reuse the Letterman Complex and many of the administrative buildings at the Main Post. The military would be financially responsible for maintaining buildings at Letterman, the Main Post, and several housing areas, thus eliminating these expenses from the Park Service budget. The Park Service would carry out natural-resource improvements similar to those of the Grand Vision, but less space would be dedicated to visitors because of military occupancy.

The draft master plan would provide the highest levels of overall resource protection and enhancement, as well as the highest levels of program diversity for the public; alternatives B and D would provide the lowest levels; alternative C lies somewhere in between. When the Grand Vision is considered in relation to the alternatives, it appears less the idealistic outcome of the planning process and more like the best of many possibilities. This is not to say that the Grand Vision is not idealistic. After all, it is a wish list of future purposes very different from the purposes of traditional national parks.

As much as the master plan embodied visionary dreams, it was also loaded with considerable economic data. In addition to providing detailed proposals for the alternative plans, the planning team felt it was imperative that the master plan provide an economic analysis not only of the Grand Vision but also of the alternative proposals. In addressing the plan's inclusion of a detailed economic analysis, Brian O'Neill explained: "We have sought to create a vision worthy of this magnificent site. Equally important, we have sought to ensure that the Presidio can fulfill its purpose as a national park in a manner that is fiscally responsible."[59] In their analysis, Park Service planners estimated that the Grand Vision as well as any alternative plan would have positive effects on the regional economy and employment (see Table 6.4).

Table 6.4
Predicted Costs and Economic Impacts of All Alternative Plans

Predicted Costs for Implementing Each Alternative

Plan	Implementation (in millions)	Annual Operating (in millions)	Annual Federal Appropriations (in millions)*
Grand Vision	$615	$44	$24
Alternative B	898	43	21
Alternative C	628	32	15
Alternative D	1,000	44	16

Predicted Economic Impact of the Presidio's Transformation by the Year 2010

Plan	Presidio Employment	Payroll** (in millions)	City Revenue (in millions)	Revenue Generated by Visitors (in millions)
Grand Vision	5,480	$127	$3.6	$35
Alternative B	7,400	175	5.0	28
Alternative C	4,100	82	3.0	28
Alternative D	7,000	100	4.0	40

Source: Compiled by author from National Park Service, *Draft General Plan Amendment: Environmental Impact Statement, Presidio of San Francisco*, NPS D-149 (San Francisco: GGNRA, October 1993: iii–ix and 82–83).

*Annual operations less the predicted revenue recovered from tenants.
**Considered to be a direct economic benefit to the city of San Francisco.

THE COMMUNITY REACTS

William Penn Mott, Jr., did not live to see the draft plan of the Presidio's Grand Vision. It is likely, however, that he would have found the project ambitious enough in many respects to "stir his blood."

When the plan was made public, it certainly stirred public reaction, in part because of a snafu. The official release date was Friday, October 16, but on October 14, the *San Francisco Chronicle,* which had received an advance copy, ran a two-page story outlining the plan. Somehow, neither San Francisco mayor Frank Jordan, nor the city supervisors, nor several members of GGNPA had been briefed on the plan beforehand. Instead, it had been leaked out in bits and pieces. Public officials who had not seen a copy of the document were annoyed that the National Park Service had not kept them informed.

Although the public unveiling was somewhat botched, in general the community of San Francisco reacted favorably, if not enthusiastically, to the concept underlying the Grand Vision. This reaction was unusual in a city known for its hyper-pluralistic politics and inability to reach consensus on any issue.[60] In fact, state and local politicians rallied quickly to support the Presidio plan. Senators Feinstein and Boxer and Representative Pelosi immediately endorsed it.[61] And, despite the premature release, Mayor Frank Jordan declared unequivocally, "we stand firmly behind the proposal."[62] Even one of the most vocal critics of the planning process, journalist Martin Espinoza, conceded, "one reason the NPS has received so much support for the unorthodox plan for the Presidio is because its vision for the park is almost universally appealing."[63] The plan was described as a fitting tribute to the Presidio, and few publicly rejected it.

This was a noteworthy achievement. Planning for the Presidio was a highly visible process, and many diverse citizens' interests were playing themselves out over the Presidio landscape—from the mushroom pickers, to the Crissy Field dog walkers, to neighborhoods concerned about traffic, to broad-scale environmental interests. The Presidio planning team faced a difficult task in reconciling the many imagined futures for the Presidio. As one observer noted, "to be able to complete a plan, and an Environmental Impact Statement, and somehow keep a coalition as diverse as that supportive of the plan is quite an achievement. There was every possibility that the planning effort could have resulted in litigation."[64] That no group or individual filed a lawsuit challenging the Presidio plan is a testament to the strong sense of community enthusiasm about the vision and concepts that had emerged during the planning process.

This is not to say that all aspects of the plan received immediate support and endorsement. Certain specific features drew fire. Several local neighborhood coalitions worried about having the University of California–San Francisco as a tenant at LAIR.[65] Others worried about the possibility of increased traffic and congestion as visitors increased in number. Much of the debate about the Presidio plan centered on the price tag rather than on the design features and goals of the proposal (aspects of the plan that are discussed in detail in the next chapter). Supporters loved the imaginative uses suggested for the park; some critics wondered how the millions needed annually for operations would be obtained.

It is worth repeating that, despite concerns about and objections to specific elements in the plan, the broad concept underlying the Grand Vision was very well supported in San Francisco. The Park Service had succeeded in developing a plan that the public would endorse.

This chapter is the beginning point from which to understand the master plan, its implementation, and the unanticipated political debate that it spawned. We turn next to a discussion of the implementation of the Grand Vision and the politics of realizing an urban national park.

CHAPTER 7

Congress and the Presidio: The Politics of Parks

> Bringing any vision to reality is always another story, involving unforeseen obstacles, clashing philosophies, sharply differing opinions of what is to be done, seemingly irreconcilable points of view, and participants of vastly different temperaments working at cross purposes. Implementing the vision will require the utmost exercise of good will, the ability to compromise, and an unshakable determination to persist in the face of hostility, discouragement and temporary defeat. We can expect plenty of figurative blood, toil, sweat and tears before the goal is reached. At the Presidio we may be about to experience both the ordeal and the unpredictable potential of a new world coming to birth.—Harold Gilliam on the Presidio plan [1]

When the Park Service unveiled the Grand Vision in October 1993, it had already achieved a significant step. But the plan was, after all, only the first in a series of steps toward the realization of the Presidio as a park.

Subsequent steps to bring the Grand Vision to reality would encounter the biggest obstacle of all: the politics of opposition. The Presidio was federal property under the jurisdiction of the Park Service, and both the management of the Presidio and fiscal support for implementation of the plan depended on the approval of the federal government. Thus the next step toward the realization of the Grand Vision was into the realm of national politics—Congress. But before the long, drawn-out, and hard-fought political struggle in Congress began, the National Park Service had to negotiate the transition period with the Army.

The Politics of Transition

The Presidio plan detailed the implementation of the Grand Vision over a period of ten years (see Figure 7.1). During the "transition phase," which began with the closure announcement in 1989, the Park Service would mobilize to provide essential services and facilities and would work closely with the Army to ensure a smooth transition. The transition phase was to be funded by a joint operating budget of $45 million per year. The Army would provide the majority of the funds, but Park Service contributions would increase as the Army phased out its operations. Federal funds would pay for necessary maintenance projects (such as upgrading water, sewerage, electrical systems, and storm drains), and the Park Service would take over public safety functions (police, fire, and emergency medical services). The park was to be under joint occupation as the Army downsized and the Park Service gradually began its programs and services.

Between 1994 and 2010, the Park Service would implement improvements and start-up programs, such as forest restoration, removal of nonhistoric buildings, working on park trails and bicycle routes, and opening many of the facilities to the public, including swimming pools, the gymnasiums,

Figure 7.1
Timeline for Implementing the Presidio Plan

1989–1993 Transition Phase	1994–1995 Beginning the Park	2000 Midway Conversion	2010 Completing the Park
Maintenance of post with joint Army/Park Service funds. Organize essential services. Park Service begins to assume responsibility for some post areas/services.	October 1, 1994: official transfer of Presidio. Begin improvements and start-up programs.	Continue improvements and refine programs.	Complete long-range improvements and set new goals.

Source: National Park Service, *Draft General Management Plan Amendment: Environmental Impact Statement, Presidio of San Francisco*, NPS D-149 (San Francisco: GGNRA, October 1993).

and the golf course. The plan set 2010 as the target date for completing park improvements and programs. This was to be the year in which the Presidio became financially self-sufficient, able to generate enough revenue from park tenants to cover all the park's operating expenses. The Park Service would then begin the task of setting new goals for the Presidio. At least, that was how the Presidio Planning and Transition Team envisioned the process. In an ominous foreshadowing of events to come, the Park Service encountered immediate challenges and political struggles.

One of the first tasks during the transition was to coordinate the integration of operations and maintenance services by the Park Service and Army. This depended on several key factors, including the Army's departure schedule, base operation issues (including maintenance), public safety, the printing and distribution of documents, the disposition of property (which included who got to keep historic items such as the Presidio's Spanish cannon), the Army's role in cleaning up hazardous waste, and the potential for continued military use of the Presidio.

In the community, concern was growing that the interim Park Service budget was insufficient and many buildings and facilities would deteriorate.[2] Major local newspapers ran stories speculating that the Army would allow the Presidio to deteriorate rather than continue to spend significant Defense Department funds on a site that would be closed within several years.[3] The *San Francisco Chronicle* reported that the Army planned to back out of a commitment to spend $10 million on repairs to roads, buildings, and electrical systems. After publication of this report, which stirred an uproar in the community and brought about considerable pressure from Representative Nancy Pelosi, the Army reversed its decision. The following week, a *Chronicle* headline announced "Army Says It Goofed—OKs Money for Presidio."[4] Nevertheless, fears over possible Army neglect persisted. In response to continued concerns, Colonel Gregg Renn, commander of the garrison at the Presidio, pledged Army cooperation, noting that "until September 30, 1994, you need not worry about the Presidio declining in terms of public safety and maintenance."[5]

Another transition period challenge was concern over environmental contamination and hazardous waste. The Base Realignment and Closure Environmental Restoration Program required the Army to clean up environmental contamination on the Presidio prior to closure. The Army carried out a series of environmental assessments of the post.[6] Although the Presidio was a clean base in comparison with many military bases, there were several landfill sites, a history of leaking tanks at the gasoline station, warehouses containing toxic wastes such as pesticides, fertilizers, and old paint, and contaminated ground water around former aircraft hangars at Crissy Field—a witches' brew of petroleum, oil, lubricants, and cleaning solvents. The studies concluded that there were several sites of environmental contamination, including approximately three hundred underground storage tanks (both active and inactive). The total cost of cleanup was estimated at $90 million. Presidio supporters were wary of generating too much publicity about the post's environmental contamination because they knew that most people would not expect a national park to contend with such a problem, and feared that this might undermine support. Indeed, many might agree that the inclusion of environmentally contaminated sites in a national park is incongruent, if not bizarre. It illustrated another characteristic of the Presidio that challenged preconceptions about national parks and environmental purity.

In a subsequent report on environmental impacts of the post's closure, the Army indicated that it might not clean up all the sites before the 1994 transfer to the Park Service, and suggested that it fence off the contaminated areas until funds were available from the Department of Defense. This proposal stirred public outrage. Letters to the editor and articles and editorials in the *San Francisco Chronicle* and *Examiner* decried the Army's record of environmental cleanup. They noted that the Department of Defense did not have an impressive record of leaving clean sites.[7] Despite the cleanup mandate of the 1989 Base Closure and Realignment Act, the local community remained wary of Army plans for remediation at the Presidio—and not without cause.

In May 1994, for instance, the Environmental Protection Agency fined the Army $556,500 for sloppy handling of hazardous wastes at the Presidio and ordered all dangerous rubbish off the post before the October 1 transfer date.[8] This was one result of a series of earlier surprise inspections of the post conducted by the EPA. EPA officials found that the Army had cleaned out several warehouses and other buildings on the post, storing hazardous wastes in a fenced area beneath an overpass leading to the Golden Gate Bridge. Although the EPA concluded there was no danger to people, local newspapers condemned the Army for faulty procedures. The Presidio's Lieutenant Colonel Steven Fredericks told local newspapers that the Army planned to remove a significant amount of hazardous waste, although some contaminated sites would not be completely cleaned before October; nonetheless, he assured the community, the Army would remain responsible.[9] In addition, the Army established a Restoration Advisory Board for Environmental Cleanup to review, comment, and advise on cleanup issues at the post. The board was to be composed of volunteer representatives from the Department of Defense, the Park Service, the EPA, and local civic agencies. All board meetings would be open to the public, and after the release of plans for specific site cleanup, the public would have a thirty-day comment period to review the proposed cleanup plans. This helped alleviate concerns that the Army, after vacating the Presidio, would ignore its responsibility for environmental cleanup.

Transition issues such as post maintenance and environmental cleanup were clearly battles about control and responsibility (financial and otherwise). The most challenging of the transition issues, however, was to come.

The Army, the Park Service, and "Operation Divot Storm"

In the summer of 1993, only weeks before the Park Service released the completed Presidio plan, the Base Closure and Realignment Commission made a stunning decision: a small contingent of the Sixth Army, some four hundred military employees, would remain at the Presidio. At first, the Park Service welcomed this news, feeling strongly that some Army presence was appropriate because the post was, after all, a national military historic landmark. Park Service officials were also ecstatic because the Army would pay $12 million per year for building maintenance and general operating costs, thereby reducing the amount of money the Park Service would need for the Presidio. The Army stressed that it was not paying rent to the Park Service. Rather, the Park Service would realize a savings by not having to maintain and repair the buildings retained by the Army.

The Base Closure and Realignment Commission recommendation had been attached to a defense appropriations bill for 1994. Embedded in the bill, in a small section toward the end, was a five-paragraph clause that prohibited any transfer of Presidio land "unless and until the Secretary of the Army determines that the parcel proposed for transfer is excess to the needs of the Army."[10] In effect, the Army could continue to control any amenities it needed to maintain its mission at the Presidio. Congress passed the legislation with little debate in November 1993.

Few paid any attention to the wording in the legislation at the time, except for representatives Nancy Pelosi, George Miller, D-Calif., and Bruce Vento, D-Minn.[11] They interpreted the clause as essentially giving the secretary of the army veto power over what the Army wanted to retain and what it would hand over to the Park Service. Together, they sent a letter to Secretary of Defense Les Aspin saying that the clause gave the Army the upper hand in negotiations and put the transfer of the post in jeopardy. The letter was never made public, but the contents eventually leaked out. Local newspapers picked up the story and published several articles. A *San Francisco Bay Guardian* editorial criticized the language of the legislation for giving the Army "an effective veto over anything the Park Service decides to do. . . . it undermines the intent of Burton's GGNRA legislation and largely renders moot the entire public planning process."[12] Similar stories appeared in other Bay Area newspapers, and the community began to debate the meaning of the clause and the Army's intentions.

In December, the Army informed the Park Service of its intention to maintain its headquarters at the historic Main Post, several hundred units of housing, the commissary, a swimming pool, the Officers' Club, the youth-service center, and the golf course. The Army planned to keep the golf course for the exclusive use of its remaining personnel and the small, elite contingent of private members who paid the Army $10,000 a year in greens fees.[13] The Army decision meant that the golf course would not be open for public use as the Grand Vision proposed.

The Army claimed that these "amenities," including the golf course and the Officers' Club, allowed the Army to "perform their mission and also provide good quality of life for their troops."[14] Army officials argued they needed the money the golf course could generate through greens fees to subsidize other on-post military activities, including events at the Officers' Club. So began a political struggle and a public debate between the Army and the Department of the Interior about who would retain control of (and hence revenues from) the golf course.

Few in San Francisco sided with the Army. Most saw the Army campaign as a threat to the ability of the Park Service to realize the Grand Vision, and they questioned the Army's need to maintain exclusive use of the golf course. The Presidio plan called for the golf course to be leased to a private operator so it could be opened to the public and greens fee revenue could help defray the costs of maintaining the rest of the park. A *San Francisco Chronicle* editorial summed up how many viewed the imbroglio:

> A mini-post manned by 350 members of the Army does not need a golf course, a huge post exchange, a commissary and exclusive use of the oldest building in San Francisco as its officers' club. The decision of the Sixth Army to maintain its headquarters at the historic fort after the bulk of troops are gone was welcomed last fall. It meant that some of the buildings that might otherwise have

fallen into idleness and neglect could be maintained properly. But this does not mean the Army should control enterprises required to provide the Presidio with revenues it will need to function. The Army will be advised to beat an orderly withdrawal from its present bargaining stance . . . or incur the wrath of local citizenry and their representatives in Congress.[15]

The issue captured public attention. Along with the deluge of editorials, letters to the editors, and other opinion articles, the popular local cartoon strip "Farley" featured a week-long series poking fun at the new Army campaign, humorously dubbed "Operation Divot Storm."

Amid public outrage and debate, the Park Service and the Army began what would become months of negotiation over what parts of the Presidio the Army would retain. Members of each "golf faction" had their own opinions about the value and the purpose of the golf course. Presidio commander Lieutenant General Glynn C. Mallory, with a handicap of 23, was reported to have said that the golf course was "vital to performing our mission"; Brian O'Neill, GGNRA superintendent, believed that "the golf course is an integral part of the fabric of Presidio history and it belongs in the park, *with access to all*" (my emphasis).[16] Interior secretary Bruce Babbitt assigned his assistant, Ira Heyman, former chancellor of the University of California at Berkeley, to negotiate "peace" between the Army and the Park Service.

The negotiations took nearly a year. When they were concluded, it was agreed that the Army would continue to operate the golf course for five years, during which public play would be phased in, reaching 50 percent by the end of the five-year period. After this transition period, the Park Service would operate the site as a public course (with a percentage of tee-times available for military use).[17] Representative Nancy Pelosi, commenting on the tempest in a teapot, said, "There will be many bumps on the road when it comes to transforming the Presidio into a park, this is just one of them."[18] Pelosi was right. This was only the first bump.

Managing the Presidio Park: A Public-Private Partnership

The success of realizing the Grand Vision depended not only on a sound plan but on a workable approach to management. During the 1980s, the Park Service had experienced budget cuts so that by 1993 it was having difficulty adequately staffing and maintaining many of its properties, let alone paying for the cost of bringing into being, maintaining, and operating what would be the most ambitious national park in urban America. The cash-strapped Park Service, GGNPA, and the Presidio Council worked together to devise a creative management structure and identify funding methods. They envisioned a coalition from all sectors of the community—economic, political, social, environmental—working to confront the many complex problems posed by the Presidio conversion. Since the centerpiece of the Grand Vision was the creation of park programs by nonpark institutions (Park Partners), the Park Service, GGNPA, and the Presidio Council realized that the management of the Presidio called for the ability to craft partnerships, secure capital investment, attract philanthropic support, lease structures, and secure income from Presidio tenants.[19] These tasks were ones in which the Park Service, admittedly, had little experience, expertise, and legal authority.[20] Park Service planners and community leaders concluded that the best form of management for the Presidio would be an innovative governing structure, a public-private partnership.

During the 1980s, the concept of public-private partnerships became increasingly popular for local development policy and planning, particularly for projects designed to revitalize old industrial cities or renew urban growth.[21] Projects undertaken by partnerships included downtown redevelopment, public housing, transportation improvements, and new sports stadiums, zoos, and city parks. As an alternative to relying solely on public initiative, these partnerships formed arrangements between private business and government in order to attract infusions of new capital. In theory, the design of the public-private partnership provides a flexible, voluntary, and cooperative alternative decision-making structure to the local or state government. Typically these partnerships result in more efficient government decision making because they are without the encumbrances of a federal bureaucracy and thus do not have the culture of caution associated with public agencies. Also, they can facilitate greater autonomy and maneuverability than state or federal agencies and bureaucracies and can be more innovative in the way they offer financial incentives, tax abatements, and project-specific subsidies. Ideally, partnerships are a collaboration between business, nonprofit organizations, and government agencies. Each partner contributes knowledge and expertise not normally available to the other partners but essential to achieving the common goal. Partnerships thus combine the best of public-minded goals with private management skills while avoiding the worst stereotypes of the "inadequate public bureaucracy" and the "predatory entrepreneur."[22] For example, individuals or groups on the business side of the partnership may contribute technical expertise, money, and equipment. And individuals or groups on the government side of the partnership help define the public benefits.

Although the concept of public-private partnerships had gained currency, there existed no widely agreed-upon procedure for organizing such an institution. In 1992, the Presidio Council contacted the consulting firm of McKinsey & Co. to research over two dozen successful public-private partnerships throughout the country and apply its findings to the specific management needs at the Presidio.[23] After detailed research, McKinsey & Co. recommended a public benefit corporation or public-private partnership as the best management option for the Presidio.[24] In theory, a public-private partnership would let the National Park Service do what it does well and supplement that expertise with the skills of professionals well qualified to manage buildings and attract and manage tenants and their respective programs at the Presidio. Importantly, the Presidio partnership would not sell or develop real estate; it would only lease buildings and sites. In this respect it was different from many of the established models of public-private partnerships in development projects.[25] The Presidio public-private partnership would "merge economic reality with park stewardship."[26] The McKinsey & Co. study concluded that this type of governing structure for the Presidio could save from 25 to 30 percent for the Presidio compared to the cost of total management by the federal government/Park Service.[27]

Many supporters of the concept, including members of the Presidio Council and GGNPA, were convinced that the Presidio partnership would make possible a new type of park governance, by replacing government bureaucracy with a more efficient, flexible, entrepreneurial structure. Many in the project, notably Jim Harvey, Presidio Council chair, understood that the Presidio would be sufficiently different from most national parks that it would need a different management approach.[28] Harvey, an articulate advocate of a broad vision for the Presidio, also recognized the growing resistance in Congress to federal funding for the national parks.[29]

With the endorsement of GGNPA and the Presidio Council, the Park Service embraced the partnership as a break with traditional park management. The Presidio would be a new model of a park, run through a public-private partnership.

The Presidio partnership was to be a collaboration of experts with diverse skills and savvy professionals experienced in dealing with market forces and knowledgeable about regional and local land-use controls (see Table 7.1). The Park Service would manage the Presidio in its areas of expertise, such as resource preservation, planning and design, visitor services, and educational programs. To deal with management challenges beyond the expertise of the Park Service, such as financing, capital improvement, programming, and leasing, the report recommended bringing in a range of experts to cut the

Table 7.1
Presidio Management: A Public-Private Partnership

National Park Service Responsibilities
- Provide overall management and ensure NPS policies are met
- Monitor compliance with the Presidio Plan and approve any modifications
- Set programmatic guidelines and goals
- Provide technical assistance, particularly in resource management
- Manage NPS facilities at the Presidio and existing infrastructure (roads/utilities)
- Provide and manage tenant and visitor services, including public safety services (such as NPS police), concessions, and interpretive services

Partnership Institution
- Manage assigned buildings and areas
- Establish partnerships with public, nonprofit and private institutions
- Negotiate and enter into leases and other contractual agreements needed to implement the plan; seek appropriate tenants for these buildings
- Fund operations by retaining and reinvesting net revenues supplemented by public and private funds
- Develop and implement public and private innovative funding approaches to help finance costs for building and infrastructure repair and rehabilitation
- Seek and accept privately donated funds

Source: National Park Service, *Draft General Management Plan Amendment: Environmental Impact Statement, Presidio of San Francisco*, NPS D-149 (San Francisco: GGNRA, October 1993: 110–112).

real estate deals, write the leases, borrow money, and run the park. The National Park Service staff would conduct the tours, interpret the history, preserve the resources, and cut the grass—traditional park ranger tasks. The partnership would combine the best of both worlds in joint responsibility.

The proposal called for the initial funding of the public-private partnership through federal appropriations. As lease revenues and income grew, the institution would become more financially self-sufficient.[30]

Since the partnership would be managing federal property (under the jurisdiction of the Park Service), it would have to be federally chartered. The Presidio partnership would need congressional authorization as a public corporation or private nonprofit foundation.[31] With these guidelines in mind, the Park Service began developing a legislative proposal to create the Presidio Corporation, a federally chartered public-private partnership. This proposal, however, would require congressional approval. In Congress, it encountered clashing philosophies and a series of political obstacles.

LEGISLATION FOR THE PRESIDIO PARTNERSHIP

There was little dispute that the base could become a premier urban park; the row was over who would pay for it and how it would be managed. Many applauded the Grand Vision but wondered whether the Park Service had the expertise to manage and administer park programs wisely. The Park Service felt it had addressed one of these concerns by endorsing an innovative management structure that would help the Presidio reach financial independence. But this proposal, however innovative, did not end the debate about how to manage the Presidio, which remained inexorably linked to the debate about how much the Presidio would cost. The annual federal price tag for the Grand Vision was estimated at $24 million for fifteen years (1994–2010), decreasing thereafter.[32] The problem was that the Presidio plan in the short run cost more than the annual budgets of Park Service units such as Yellowstone (annual budget of $17 million) and Yosemite (annual budget of $15 million). This meant the cost of the Presidio plan was vulnerable to a barrage of criticism. In fact, several involved with the Presidio later noted that the Park Service was never able to convince Congress and the congressional staff that the Grand Vision was economically viable.[33] Part of the problem occurred because the master plan documented an implementation cost of $660 million, a staggering amount for any political elite to consider for a park. Even though the Park Service took great pains to spread the $660 million figure over twenty-five years, members of congress and their staff never got past the number $660 million, which would haunt the Presidio Trust legislation.

Representative Nancy Pelosi announced she would sponsor legislation to create a public benefit corporation for the Presidio.[34] Pelosi had this to say about the concept of a Presidio public-private partnership:

> The Presidio is an example of a base closure that will save taxpayers' dollars while serving an enhanced national purpose. Over time, project costs for the Presidio will be less as a national park than as a military facility. In the current fiscal environment, it is essential to look for innovative ways to manage our federal assets that save money while ensuring accountability.[35]

In October 1993, the House Subcommittee on National Parks, Forests, and Public Lands held its first hearing to consider the draft Presidio legislation. In addition to Park Service officials, numerous supporters came forth to endorse the Presidio plan and to advocate the creation of a federally chartered public benefit corporation. Advocates included Toby Rosenblatt (chair of GGNPA and vice chair of the Presidio Council), William Reilly (former administrator of the U.S. Environmental Protection Agency and senior adviser to the Presidio Council), Amy Meyer (member of the GGNRA Advisory Committee and "matriarch of the GGNRA"), and Michael Alexander (chair of the Sierra Club's Presidio task force). Toby Rosenblatt explained the need to reconcile the Grand Vision with a pragmatic approach to its implementation: "to be successful in this era of fiscal restraint, the Presidio's buildings and related facilities can and must be managed to ensure maximum responsiveness to public needs with minimum costs."[36]

Although supporters were numerous, it was the Presidio Council that most diligently and effectively lobbied for the Presidio Corporation legislation.[37] It would also continue its leadership role in soliciting philanthropic contributions to the Presidio. Many in Congress considered philanthropic support an important element in the Presidio plan because it gave credibility to the assertion that the Grand Vision truly had local and even national support. After all, it is one thing to publicly endorse the idea of the Grand Vision; it is another thing to donate money for its realization. Many on the council felt confident that the Presidio could raise at least $30 million in philanthropic donations over a fifteen-year period. As William Reilly noted:

> The philanthropic community has already provided some $2 million in support of pro bono work for the Arthur Andersen study, for the McKinsey study, for a number of services and activities that have supported the Presidio. $30 million, which is anticipated over the next 15 years or so, is in my view a quite conservative number. . . . The community has given over the last few years some $75 million for the new museum of modern art, $30 million for the main library extension, another $18 million for the San Francisco Zoo, and I do not know how much for the Davies Symphony Hall. There is an extraordinarily generous community of people there who are waiting and in some ways becoming impatient for a management structure for an ability to contribute real and permanent improvements of Crissy Field and of other facilities in that park.[38]

In supporting the proposed Presidio Corporation legislation, William Reilly also testified:

> The Presidio, it is true, presents unconventional problems and would be a different kind of park. . . . Some of you may question whether the "vision for the Presidio" is practical, given the climate of fiscal restraint that we are currently faced with. With the correct management and an appropriate financing plan,

> I am convinced that this park can be established at a cost that is in line with that of other national parks. We must establish a new management partnership that can manage building properties with expertise and efficiency. The Administration has signaled its intention to "reinvent" government. Perhaps the best way to do this is to create partnerships between government and private organizations—partnerships that work by combining private sector efficiencies and expertise with public sector goals and objectives. The Presidio is a perfect place for such a partnership.[39]

In November, Pelosi officially introduced the bill, H.R. 3433, which established the Presidio Corporation. Shortly thereafter, Senator Boxer and Senator Feinstein introduced a companion bill, S. 1549, to Pelosi's Presidio Corporation bill. Said Senator Boxer:

> The Presidio can truly be a model of successful military base conversion. Every other national park unit has a worrisome and growing backlog of deferred maintenance and capital investment. By contrast, this legislation gives the Presidio a managerial and financial structure which decreases the park's backlog year by year, and does so by using revenues generated by the Presidio.[40]

Supporters at first thought Congress would move quickly on the bill. According to the anticipated schedule, the bill would be approved by the subcommittee in February, passed by the House and Senate in March, and signed by President Clinton in April 1994—six months before the official transfer date; and the Presidio Corporation could be organized and operating by 1995, only a few months after the transfer. But few things in Congress ever keep to schedule, and the Presidio plan encountered more challenges.

PRESIDIO INC. OR PRESIDIO TRUST?

In January 1994, the Presidio Council met and agreed that the name Presidio Corporation did not convey the right image because it connoted commercialism rather than stewardship. The council suggested changing the name and offered several alternatives, including Presidio Trust, Presidio Park Foundation, and Presidio Park Conservancy. The Department of the Interior agreed with the Presidio Council. In its review of H.R. 3433, the Interior Department recommended that "the public benefit corporation should be renamed the 'Presidio Trust' . . . *this change is of symbolic value* and should help to assuage concerns that the entity will be linked somehow to business and development interests" (my emphasis).⁴¹ In March, the name was changed to Presidio Trust.

In part, the Presidio Council had responded to a series of critical articles in the *San Francisco Bay Guardian* and elsewhere.⁴² One article asserted that the Presidio Corporation would turn the Presidio into a national park run more like a "national business park," and predicted that the Presidio Corporation would be nothing more than a money-making machine, a tax-free corporation operating like a super-redevelopment agency.⁴³ Another article said the Park Service's primary goal was really to maximize revenues from leasing Presidio properties and criticized the agency for surrendering most of its authority to develop and run the new Presidio park. The writer believed that the Presidio would become a government-subsidized business park, that the Presidio would be marketed like a commercial shopping mall, and that developers would be able to avoid the city's strict controls on growth and zoning regulations.⁴⁴ Another reporter warned that the Presidio Corporation would be dominated by prodevelopment forces who would be tempted to lease Presidio space to the highest bidder, thereby obscuring the Grand Vision of the environmental education center. Two journalists, Martin Espinoza and Joel Ventresca, were vigilant in their outspoken criticism and labeled the Presidio partnership "Presidio Inc.," suggesting that the partnership would rely on developers, big business, and large institutions to plan and operate the park.⁴⁵ They urged Pelosi to make several changes to the legislation to ensure more public accountability and public access to the workings of the partnership. The *Bay Guardian* was not alone in its worry. The National Parks and Conservation Association was wary about the Presidio Trust, arguing that the national parks should be fully funded by the federal government.

Concern over the probable presence of private business at the Presidio and the Presidio Trust partnership highlighted fundamental ideological differences about how parks should be administered. Some argued that the presence of any private interest at the Presidio would degrade national park values.⁴⁶ For many of the Presidio park advocates, however, the most important task was to *preserve the integrity of the Presidio as an urban open space*. They felt that because of the number of buildings and the complexity of the site, there was no viable alternative, and they argued that the idea that only nonprofit organizations reside at the Presidio was narrow-minded and would lessen the chances of the Presidio surviving as a park.

GGNPA executive director Greg Moore believed that some of the local resistance to the Presidio Trust arose because of the way in which the Presidio's planning had occurred. During much of the visioning phase and even in the final plan, the public had been encouraged to dream big. When the

Grand Vision was published, and the public heartily endorsed the plan, most people assumed that the vision would come true and the future Presidio would resemble a nonprofit campus for the exploration of environmental issues. The plan was enthusiastic about the possibilities of a nonprofit global center but only vaguely mentioned ways in which private business would occupy buildings on the Presidio.

Pragmatists (like Jim Harvey and members of the Presidio Council) realized that the creation of an organization akin to the Presidio Trust was the only way to ensure congressional approval. But as Congress began to raise issues of fiscal control and responsibility in the initial hearings, the Presidio Trust became a difficult public relations problem.[47]

In April 1994, Pelosi responded to the criticism raised by the *Bay Guardian* articles, as well as to neighborhood concerns about the potential impact of large-scale events at the Presidio. She proposed to amend the legislation to make it clear that the Presidio Trust would be "sensitive to the unique needs and local impact of the Presidio as an urban national park."[48] Pelosi added a preamble to the bill stating that the purpose and intent of the Presidio Trust would require that its meetings be held at the Presidio and be open to the public. Another amendment mandated that the trust's financial records would be audited annually to ensure public accountability.

Despite these changes, reporters Ventresca and Espinoza continued their virulent attacks against the Presidio Trust. At first, many of their criticisms were constructive. But even after Pelosi incorporated many of their suggestions into the legislation, they continued to paint the Presidio Trust as a sinister plot to commercialize the Presidio. Throughout the legislative process, the *Bay Guardian* continued its tirade; but somewhere along the way, its criticism evolved into cynicism, and its tirade disintegrated into obsessive, irrational whining. The paper was unwilling to accept the fact that some kind of public-private partnership was necessary in order for Congress to accept the Presidio plan. And, though adept at criticism, the *Bay Guardian* did not propose a single alternative to the Presidio Trust. This fact weakens the paper's claim to be an advocate.

Politics in San Francisco had altered but not derailed the Presidio legislation.[49]

A Tale of Two Bills: Political Opposition in Congress

Success in Congress often depends on quick action while the momentum is in your favor. Unfortunately for the Presidio, the potentially friendly Congress was about to change as a new political climate emerged. Many members asserted a renewed interest in balancing the federal budget, and Congress was becoming more fiscally conservative. For 1993, outgoing President Bush had approved $14.8 million for the National Park Service to spend at the Presidio to help facilitate the transition. The Park Service requested $25.4 million for the 1994 budget, a $10.6 million increase over the 1993 appropriations. The Park Service needed the increase because it was taking control over more areas of the Presidio from the Army.

As the Park Service was putting the finishing touches on the Presidio plan during the summer of 1993, the proposal for the 1994 budget went forth in Congress. Republican representative John Duncan of Tennessee sponsored a legislative amendment to the 1994 Interior Department appropriations bill. Duncan's proposed amendment would *reduce* national park system funding by the exact amount President Bush had authorized a year earlier for the Presidio. The Duncan Amendment, as it became known, also intended to freeze funding for Park Service operating expenses at the Presidio.[50] Duncan saw Presidio funding as a litmus test of congressional resolve to tackle the federal budget deficit. A letter from Duncan and several proponents of his legislation outlined the objections to the Presidio park:

> While we support the inclusion of certain land located in the Presidio within the National Park System, we cannot support the setting aside of the entire 1,480-acre area for national park purposes. It is our opinion that the National Park Service should not be in the business of operating or leasing bowling alleys, movie theaters, golf courses, hospitals, housing, medical facilities and daycare centers. . . . This park will be the most expensive national park in the country. It will drain money from every other park in the Nation. We believe that with a federal debt of over $4 trillion dollars,[51] and a current shortfall of several billion dollars for operations, facilities construction and land acquisition at existing units of the park system, any further expenditures of taxpayers' dollars on this very expensive project is ill-advised.[52]

Along with the letter, Duncan included a photograph of the Presidio Pet Cemetery with a headline asking, "Is This Your Vision of a National Park?"[53] Duncan wrote that in order to offset the cash outlay, the National Park Service was

> examining tenant proposals from: Artist Colony and Homeless Housing, Pickle Family Circus School, Golden Gate Bungee Tower, Frisbee Golf Course, Surf Condition Hotline, and the Institute for Public Golf, to name but a few. . . . If

you agree with us that Federal subsidies for such activities in a time of budget crisis are ridiculous, then please support our amendment to freeze National Park Service funding for the Presidio.[54]

Although the Duncan publicity about the Park Service plan was astoundingly inaccurate in its portrayal of the planning process and ridiculed the Presidio plan as nothing more than a compilation of quirky proposals, the letter received the endorsement of many Republican congressional representatives. To combat the Duncan proposal, Nancy Pelosi teamed with Bruce Vento, D-Minn., chair of the national parks subcommittee, and George Miller, D-Calif., chair of the natural resources committee. In testimony before the House, Pelosi argued persuasively:

> This is perhaps one of the most exciting base closure conversion projects in the country. . . . The Presidio plan, we hope, will be a model: it will be a source of jobs, it will preserve the beauty of the Presidio and the GGNRA, and it will, yes, indeed, it will produce revenue to help reduce the deficit. Every effort is being made to maximize revenues to the park and minimize the impact on the Federal Treasury. I believe we can look forward to an exemplary national park that will continue to benefit millions of worldwide visitors, bringing revenues to our park, and that many generations will walk its paths after us.[55]

The campaign to defeat the Duncan Amendment was successful. On July 15, it was defeated by a vote of 230 to 193 in the House, a difference of only 37 votes. The amendment, however, stalled the progress of the Presidio Trust legislation by creating significant congressional resistance. Presidio advocates lost critical time. Some felt that a negative side effect of the Duncan defeat was that the high cost of the Presidio plan became well publicized. In the early days, only supporters of the Presidio knew of the $660 million figure; now everyone seemed to know.[56] This made the Presidio even more vulnerable to attack. It became apparent that what had been considered a "done deal" by many was anything but. This attempt to derail the Presidio plan was only the first of many.

Almost as soon as the Park Service released the Grand Vision, House Republicans challenged the plan for the Presidio, questioning the logic of spending hundreds of millions of dollars at a time when existing national parks were falling apart. Jim Coon, spokesman for Representative Duncan, argued "the reason we're shutting down military bases across the country is because of our enormous debt. The fact of the matter is, the National Park Service can't afford to run the parks they have right now."[51] At the October 1993 hearing, during which Representative Pelosi introduced her plans to sponsor legislation creating the Presidio public-private partnership, others took the opportunity to criticize the Park Service plan. Representative James Hansen, R-Utah, stated, "I'm hearing from park rangers how in disrepair the parks are, how bad they are, how we can't keep them up and you folks come in with all kinds of enthusiasm to spend additional money?"[58] Hansen and others alleged that the Presidio was a local park, not a national resource. But Presidio proponents, notably several Presidio Council members who attended the hearings, effectively communicated the value of the Presidio as a national resource.[59] Many other Presidio Council members wrote letters to key congressional members, asking them to publicly support the creation of legislation for the Presidio. In fact, the presence of several Presidio Council members at the hearing helped refute the allegation that the Presidio was merely a local park by proving that a national constituency existed for the Presidio.[60]

The criticism over the cost of the Presidio and the condition of other Park Service property was a reminder that the Grand Vision, however grand and worthy, would provoke opposition. In commenting on the complexities of turning the Presidio into a national park, an exasperated Interior Secretary Bruce Babbitt said, "I am wearily confident that we are really going to keep this thing on track."[61] He bemoaned the political mine fields that had to be traversed as politicians fought over the details of the plan, explaining, "I've learned that there are 2 million experts on the future of the Presidio."[62]

As exasperating as these preliminary challenges were, they helped mobilize GGNPA and the Presidio Council. Together these nonprofit organizations worked to formulate effective strategies for constructing a lobbying effort, dealing with the political challenges to the Presidio plan, and arguing against proposals like that of Congressman Duncan.[63] Park Service official Brian O'Neill acknowledged that "the Park Service itself is not in a position to mobilize and lobby. We have to depend upon the community."[64] The expertise and the commitment of Presidio Council members and GGNPA were instrumental in gaining support for the Presidio Trust legislation. Presidio Council staff spent endless hours preparing briefs and memos with which council members would lobby senators and representatives. GGNPA directors and Presidio Council members drew up a list of potential avenues of attack from opponents and prepared convincing counterarguments. The Presidio Council and GGNPA led the lobbying effort; without their political savvy and dedication, it is likely that the Presidio Trust legislation would have withered away in subcommittees. Brian O'Neill recalled:

> The presence of the Presidio Council made all the difference. They gave a national context to the project and the level of obligation and enthusiasm was astonishing. We get paid in the Park Service to market the value of the National Parks. It seems so self-serving when we do it. But when you get a group of business leaders and a group of people who have distinguished themselves in the academic world, or the environmental arena or the social service arena, and they are saying the Presidio represents a profound opportunity to create a park . . . well they have credibility. The sheer magnitude of this project would have scared off a lot of people without the wisdom and endorsement from the Council. I really credit the Presidio Council for being an extremely important component in educating Congress and getting their support.[65]

Duncan's stance against the Presidio as a park was based on a different interpretation of the 1972 GGNRA legislation. He argued that the Presidio's inclusion in the GGNRA legislation was vague and incidental, and in one regard he was correct. The idea of turning the Presidio into a park originated in a single obscure sentence in the 1972 legislation creating the GGNRA. Whereas some saw the vague statement as a possibility to dream "Grand Visions," Duncan argued that because the incorporation was incidental and the military had no plans to leave the post at the time, Congress had not fully reviewed the concept of taking over the entire Presidio. Therefore, Duncan concluded, the best solution was for the federal government to retain only some of the Presidio lands and divest itself of most of the buildings and other facilities. What Duncan failed to realize was that although the inclusion of the Presidio was vague, it was not incidental. As Chapter 3 revealed, it was a shrewd and farsighted political maneuver by the legislation's sponsor, Philip Burton.

But this did not stop Duncan from proposing his own Presidio bill. In January 1994, he penned legislation to drastically change the Presidio. His bill, H.R. 4078, would "amend the Act establishing the Golden Gate National Recreation Area," in effect altering the Presidio clause in the original GGNRA legislation.[66] Duncan's bill proposed eliminating more than 1,200 acres, or 85 percent of the Presidio, including the golf course, Letterman/LAIR, the Main Post, Fort Scott, all the housing areas, and the Presidio Forest/Presidio Hill. These areas would be sold to private developers; profits from their sale would be funneled to the partnership to manage the remaining areas—the coastal bluffs and Crissy Field (see Figure 7.2).

The Park Service and the San Francisco community believed this bill would "carve up and sell to the highest bidder" the "heart and soul of the Pre-

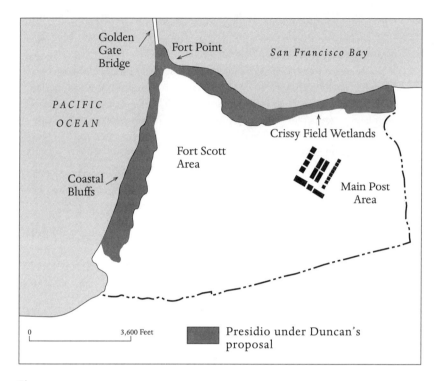

Figure 7.2
The Presidio According to Duncan. Tennessee representative John Duncan's legislation, H.R. 4078, detailed the proposed boundaries of Park Service land. The remaining areas of the Presidio, more than 1,200 acres, would be sold to private developers in order to raise capital. Notice that only the "nature" areas of the Presidio would remain with the Park Service and the "culture" areas, such as many historic buildings, would be excluded. *Source: Map by author, based on description of proposed Presidio lands in H.R. 4078.*

sidio." "The Duncan proposal would be the death knell to our plans and hopes for a great national park," summed up Pelosi.[67] Luckily for Presidio supporters, Representative Duncan and other opponents were unaware of the hazardous waste contamination on the post (or chose not to use this as ammunition in the debate). Imagine the hostility and political resistance that might have been generated had Duncan mentioned that certain areas of the Presidio could qualify as an EPA "Superfund Site!" It is hard to determine what impact this information could have had given widely held ideas that national parks should be environmentally pristine, but it is possible that it may have swayed some of the undecided members of Congress to support Duncan's Presidio bill.

In addition to challenging the Presidio's integrity, Duncan's bill posed a much more fundamental threat. The Presidio, as stipulated by the 1972 legislation, was technically already National Park Service property; it was only awaiting full incorporation when the Army declared it surplus. As William Reilly explained, "our challenge now is not one of deciding whether the Presidio should be a national park—it already is. Our objective must now be to manage this park in a manner that protects it at the least possible cost to the taxpayer."[68] Duncan's bill would in effect remove property from Park Service jurisdiction, and the implications of such an action were great. Brian O'Neill, GGNRA superintendent, told one newspaper, "the Duncan proposal has serious ramifications for the entire park system. If we begin to get into a process where individual members of Congress, *for economic reasons,* are proposing the removal of units or portions of national parks, the national park system, as we know it today, is going to change" (my emphasis).[69] Michael Alexander, of the Sierra Club's Presidio task force, called Duncan's version of a public-private partnership for the Presidio the "Duncan High-Rise Condominium Development Corporation" and believed Duncan's legislation "would leave all parks vulnerable to development."[70] Supporters tried to recenter the debate and argued the focus should be not on whether the Presidio deserved to be in the Park Service but on how to empower the Park Service to manage and operate its newest property.

The wider implications of Duncan's bill were overshadowed by his excessive attention on the cost of the Presidio. Duncan argued that his legislation was all about money, not about fundamental changes to National Park Service property. He claimed that "estimates show it will cost $1 billion to acquire the Presidio as a park and $45 million per year to run it."[71] Pelosi disagreed and challenged Duncan's figures:

> The Duncan bill is a waste of time. It's disguised as a discussion about money, which we win anyway, because his numbers have been debunked.[72] But this isn't about numbers, this is about open space. We want the Presidio to become a park and they want to sell it off and privatize it. We have a vision that is environmental and global in scope, not one driven by development.[73]

Pelosi was correct in one respect. Duncan's appeal rested on inaccurate figures for the Presidio's transformation. As William Reilly commented: "Certain myths have been circulating about the Presidio—myths that tend to make our task seem more daunting than it really is—myths that overstate square footage at the Presidio, overestimate capital improvement costs and operating expenses."[74] During the planning process the Presidio Council had hired Keyser Marston Associates, a San Francisco firm specializing in real estate development, to analyze the Park Service plan and review the costs associated with implementing the Grand Vision. Keyser Marston Associates concluded that the cost of implementing the Presidio plan was lower than the Park Service had originally estimated and much lower than the $1 billion figure Duncan cited (see Table 7.2).

In addition to informing congressional members about the revised estimate of Presidio expenses, Pelosi argued that it might, in the end, cost more to sell off the land and guide the property through various zoning and regulatory processes than it would to pass her bill calling for lease arrangements through the Presidio Trust.[75] As it turned out, this was an extremely important argument. Presidio advocates argued that if Congress endorsed Duncan's bill and sold off parts of the Presidio, they would not realize any savings because the city of San Francisco had zoned the Presidio "urban open space."

Table 7.2
Presidio Cost Analysis

Presidio Total Operating Costs
  1994–2005 $28 million per year
  2005–2008 $22 million per year
  2008– $16 million per year

Presidio Capital Costs
  1994–2010 $590 million total*

  $345 million financed through the Presidio Trust (example: lease of building space)
  $245 million funded through a combination of federal appropriations and philanthropy
  • $30 million over 15 years from philanthropic sources
  • Remainder from federal funds

Source: Keyser Marston Associates, *Presidio Analysis* (San Francisco: Keyser Marston Associates, 1993).

*Original Park Service estimates were $666 million. However, experts concluded that this could be reduced to $590 million for two reasons. First, the Army planned to remain at the post and would incur some maintenance expenses. Second, the Presidio Trust structure would be more efficient than a government agency.

Experts agreed that because of the many park advocates, any attempt to rezone the Presidio would meet with a barrage of litigation. Some believed that lawsuits could take ten or fifteen years to settle before the city would be forced to rezone the area. In the meantime, few potential investors would be willing to put money into a project before the area was rezoned. Although selling off the Presidio may have seemed a viable solution, the federal government would have had to pay for the maintenance and upkeep of the Presidio during the time it took to rezone the space; otherwise, the buildings and infrastructure would have deteriorated to such a degree that selling them would be difficult. GGNPA and the Presidio Council argued that the fiscally smart thing to do would be to set up a mechanism by which the buildings would be leased and the property maintained—something similar to the Presidio Trust. Once the reasonableness of this idea became evident to many in Congress (gathering the necessary information did take some time), the knee-jerk "let's sell it" reaction dissipated, and the Presidio Trust seemed to offer the best way to handle the property.[76]

Moreover, Duncan's argument for selling off nearly all of the Presidio ignored the comparative costs of maintaining the Presidio. The average annual operating budget of the Army at the Presidio from 1986 to 1992 was $60 million.[77] The Park Service planned budget of approximately $25 million per year represented one-third of the federal allocations appropriated to the Army. Senator Feinstein argued that "less federal dollars will be spent at the Presidio this year than any time in the recent past, and will continue to decline as lease revenues are generated."[78] Senator Barbara Boxer added, "we are saving money from what the Army used to pay for the Presidio; $60 million a year is what it cost to operate that base, and that is 264 percent more than will be needed to run the park."[79] While Duncan decried spending millions on a park, he apparently had had no qualms about approving defense appropriations of twice that sum for the same space. However, the fact that the Presidio as a national park was significantly less expensive than the Presidio as an Army post was, interestingly, of little consequence in the national political debate.[80]

Despite convincing and provocative arguments in favor of the Presidio plan, the Presidio Trust legislation continued to be enmeshed in partisan politics. Duncan's arguments, however partial and inaccurate, were persuasive. By April 1994, his legislation had seventy cosponsors, nearly all Republicans. Congress would choose between the two legislative proposals, thereby determining the future of the Presidio.

Mobilizing the Community for the Presidio Trust

In the mobilization and lobbying effort to defeat the Duncan legislation and ensure passage of the Pelosi bill, much of the credit goes to Pelosi's staff, especially legislative aide Judy Lemons, as well as to the Presidio Council members and staff and GGNPA. In Washington, council members and GGNPA staff diligently worked the steps of the Capitol to win support for the Presidio. In the Bay Area, numerous organizations helped to publicize support for the Presidio Trust legislation. On Earth Day 1994, the National Parks and Conservation Association organized a "March for Parks" in San Francisco to show support for the Presidio. Despite winds and cold temperatures, more than fifteen hundred marched in support of the Grand Vision. "This sends a clear sign that the community supports maintaining this as a national park. We will win the fight in DC," Pelosi promised the crowd gathered at Crissy Field.[81] Special guest speaker Republican governor Pete Wilson endorsed the Democrat Pelosi's bill and opposed the Republican-backed Duncan legislation.[82] In California, at least, the Presidio appeared to have bipartisan support.

By May, the community had begun to rally behind Pelosi's bill in earnest. An editorial in the *San Francisco Examiner* warned the Presidio "needs to be saved from the misbegotten proposal of Rep. John Duncan, R-Tenn., to sell off all but 200 acres to private buyers. With the likelihood of producing a costly stalemate resulting in mothballing and decay of buildings, the Duncan plan would not even save the government money while squandering an irreplaceable national treasure. Save the Presidio. Go with Pelosi."[83] Another editorial made an impassioned plea:

> Just imagine the outcry if a politician proposed selling 90% of the Golden Gate Park to the highest bidder. This is exactly the kind of commercialization that will take place in the Presidio, San Francisco's Other Great National Treasure, if Duncan gets his way. From the vantage point of a budget-strapped Washington politician, Duncan's plan to raise cash by selling the Presidio as ocean-front real estate makes sense to an alarming number of Congressmen. . . . In 1993 Duncan's bill came within 37 votes of passage. This year Duncan has 60 co-sponsors for his plan to sell off the heart and soul of the Presidio. We must present a strong unified front and tell Duncan, and his allies, that the Bay Area won't sit idly while they try to plunder the Presidio.[84]

Even the often critical *San Francisco Independent,* which had criticized Pelosi's legislation for lacking public accountability, published an editorial noting that

Pelosi had responded to these concerns and proposed modifications to her own legislation. The editorial stated, "There is no question that Pelosi's legislation is the best hope for saving the Presidio. But the Presidio bill needs help. Pelosi needs to let Congress know that the people are behind her. We urge all San Franciscans to write, fax, or phone Pelosi's office to express that support."[85] Mayor Frank Jordan also supported the proposed legislation and attended the Senate hearings, stating emphatically that the city was committed to the successful implementation of the Presidio plan.[86] The San Francisco Board of Supervisors unanimously passed a resolution supporting the Presidio Trust bill. So too did the Neighborhood Association for Presidio Planning, an umbrella group consisting of the neighborhood associations that abut the Presidio and would be most affected by the Presidio's transformation. Numerous organizations endorsed the legislation and sent letters of support to Pelosi, including the Environmental Defense Fund, the Sierra Club, the San Francisco Chamber of Commerce, and the American Institute of Architects. Pelosi took with her to Washington the support of more than thirty local and national organizations.

Showdown at the Capitol

Hearings for the debate about the Presidio's future were scheduled for May in both the House and the Senate. Surprisingly, the Duncan bill fizzled at the first congressional hearings on May 10. Duncan did not show up and did not send a representative to defend his bill. In contrast, more than a dozen people testified in support of Pelosi's bill, including Senator Dianne Feinstein and Senator Barbara Boxer; Roger Kennedy, director of the National Park Service; Frank Jordan, mayor of San Francisco; and William Reilly, senior adviser to the Presidio Council. As a result, the four-hour hearing focused on the merits of Pelosi's bill and discussed ways to ensure financial self-sufficiency and protection against the commercialization of the park. The Presidio Trust legislation was modified to cap annual federal appropriations on the Presidio at $25 million, thereby allaying fears that it would become a money pit.

On May 12, the hearings moved to the Senate subcommittee. "With the competing demands for what little (federal) money we have, when you consider that this park for the next 10 years will cost as much to operate as Yosemite or Yellowstone, either one, a lot of people have questions," began Senator Dale Bumpers, D-Ark., chair of the national parks subcommittee of the Energy and Natural Resources Committee.[87] Senators Feinstein and

Boxer reminded the subcommittee that the federal government would save millions by converting the military base into a park. Despite tough questions, the subcommittee approved the legislation, and it moved to the full committee for a vote.

In June 1994, the Energy and Natural Resources Committee approved H.R. 3433 by a vote of 28 to 14 (with all 28 Democrats supporting the bill and all 14 Republicans opposing it). On August 18, 1994, the full House of Representatives approved H.R. 3433. Another major hurdle had been overcome.

Presidio supporters expected the Senate version to be passed by September or October. But the months came and went without passage while Senate Republicans filibustered the bill. Then, changes on the national political stage transformed Congress and thus the debate about the Presidio.

The fall 1994 national election swept into office supporters of Newt Gingrich's "Contract with America" and dramatically altered the control of the House of Representatives. The Republicans assumed control of the House and thus controlled appointments to committees and subcommittees. The supportive Democrat Bruce Vento was no longer chair of the House Subcommittee on National Parks, Forests, and Public Lands. In his place was James Hansen, R-Utah, an earlier critic of the Presidio plan. The new Senate threatened a filibuster on the bill and 1994 ended without legislation for the Presidio Trust.[88] The trust had become a legislative casualty.

Meanwhile, from Post to Park

While Congress stalled on the legislation for the Presidio, the Army vacated and the Park Service celebrated the transfer of the post from "olive-drab to park-green." On September 30, 1994, at 4 P.M. near the flag pole at Pershing Square located on the Main Post, the ceremony to deactivate the Army's Presidio Garrison and the Headquarters Command Battalion began. Presidio Army troops paraded by their senior general, the cannon sounded, the American flag was lowered, and the Sixth Army's colors were retired for the last time. It was a solemn retreat ceremony full of military pomp, patriotism, nostalgia, and pride. In the valedictory speech, Lieutenant General Glynn Mallory, commanding officer of the Sixth Army, reminded the crowd of the special meaning the Presidio had in Army culture: "It was said that an officer in the old Army had three ambitions: to make colonel, to be assigned to the Presidio, and to go to heaven."[89]

Following the furling of the colors, melancholy Army officers and their families adjourned to a tent nearby. Others in the crowd watched the sun set over the Golden Gate Bridge and walked to another nearby tent for another kind of gathering—the Golden Gate National Park Association benefit celebration. Hundreds of people gathered on this eve of the transfer, attired in "Presidio Festive," to eat, drink, be merry, and await the stroke of midnight (see Figure 7.3).

At midnight, after 218 years as a military installation, the Presidio became a national park. Its new history began.

The ceremony, of course, was more symbolic than real. The next morning, the Army still occupied about one-third of the Presidio's buildings. The Presidio management plan was still under debate; funds for the Park Service

Figure 7.3
From Post to Park: A Gala Celebration. San Francisco artist Michael Schwab created this symbol for the Presidio's transfer. He said he was inspired by Iwo Jima to convey the power in the flag. In this instance, he used an Army soldier and a park ranger to highlight the gesture of the transfer from Army post to national park. Schwab's logo was used in numerous Park Service publications. *Source: The Golden Gate National Parks Association.*

were under constant threat. The post looked pretty much as it had the day before, but the Presidio nonetheless had become part of the GGNRA. Sometime between taps and sunrise, signs at the Presidio gates that had said, "You are entering a military reservation . . . cars are subject to search," were replaced with "Welcome to the Presidio of San Francisco, Golden Gate National Recreation Area."

On Saturday, October 1, the National Park Service hosted a series of free public events to celebrate the new park. Various activities, from walks to tours to films, and historical reenactments were scheduled throughout the day. The transfer celebration occurred while Congress fought over how (and whether) to fund the Grand Vision.

BACK TO CONGRESS

Representative Pelosi reintroduced the Presidio Trust legislation, as H.R. 1296.[90] The legislation again began the long, contentious journey through subcommittees and full committees. But things were very different in the 104th Congress.

In early 1995, with Republicans in control of the House, the Presidio Trust legislation faced a difficult challenge. Many Presidio advocates were worried. The votes on the various Duncan proposals had been party-line votes. If members continued to vote along party lines, the Presidio was doomed. Pelosi, however, was determined and worked to build bridges with the new Republican leadership. She was aided by the fact that the Park Service, GGNPA, and Presidio Council had worked with many members of the Republican staff on the previous committee—many of those staffers were now in charge of the national parks subcommittee. This meant Presidio advocates did not have to educate new staff and congressional members about the Presidio. Pelosi worked with the new leaders from the outset to fashion an agreement to preserve the Presidio as a national park. Although she realized some compromises would be necessary, she was committed to ensuring the integrity of the Presidio.

A new movement, however, again threatened the Presidio's inclusion in the GGNRA. Allen Freemeyer, staff director of the House subcommittee that oversees the National Park Service, expressed his views on the Presidio: "There is clearly a large part of that base that should not be in the national park system. It's a lemon that's been handed to us and we'll have to do something about it."[91] Republicans in Congress began to talk about a "National Park Closure Commission" to do for national parks what the Base Closure and Realignment Commission had done for military bases. And they hinted that the Presidio would be at the top of the hit list. This proposed legislation became known as the Hefley Bill. Robert Chandler, the National Park Service's Presidio general manager, succinctly summed up the political turn of events: "Our future is very uncertain."[92]

Things went from bad to worse. In December 1994, less than two months after the Presidio's transfer to the Park Service, the Army, which had planned to keep a small number of personnel at the post (and hence provide $12 million a year in operating costs), announced it would be pulling out of the Presidio altogether. The Pentagon had decided to eliminate the entire Sixth Army as part of the ongoing defense downsizing.[93]

In the same month, the University of California at San Francisco, which had been negotiating for several years to become the sole tenant at the Letterman Complex, decided not to exercise its lease. The Presidio thus lost a prospective anchor tenant and much-anticipated revenue. Many criticized the Park Service for its lack of negotiating skill.

The Park Service was caught in a bind. The Presidio Trust had not been authorized, so the Park Service, by default, was the interim negotiator (with admittedly little expertise and even less success). Moreover, the Park Service could not borrow money to fix up buildings (only the Presidio Trust could do this), which meant it had to attract tenants to empty, dilapidated structures. The Grand Vision appeared in jeopardy.

In Washington, Pelosi and her legislative aides spearheaded a renewed campaign for the Presidio, along with Boxer and Feinstein and their respective staffs. This effort included arranging for key members of Congress to tour the Presidio. These tours proved invaluable. Many who visited became convinced that there was little question of the Presidio's significance, and more than a few changed their initial opposition to the legislation.[94] Slowly, Presidio advocates turned the corner on the Presidio Trust legislation. The rhetoric changed dramatically. In late 1993, the rhetoric had been "We need to get rid of this and parcel it off"; by early 1995, the rhetoric was "How can this be put together in a way we can afford?"[95] In September 1995, the House passed Pelosi's bill to create the Presidio Trust. To many, the House vote for the Presidio Trust by a 3 to 1 margin was remarkable, given the changed character of the House leadership.

However, to secure that bipartisan support, advocates were forced to make several concessions. First, Presidio advocates were troubled by a new "Reversion Clause," which said that if the trust failed to achieve a sufficient degree of financial self-sufficiency, the property would revert to the Defense

Department for disposal. The criteria for self-sufficiency and failure were unclear. In the early years of planning, no one had anticipated that there would be a requirement for *no* federal appropriations.[96]

Second, Congress removed even the management of the open spaces from the Park Service. The entire Presidio, not just the buildings, would be under the purview of the Presidio Trust. This was a clear departure from the concept of partnership at the Presidio. This concession was a reflection of a changed Congress, which in the course of two years had inserted increasingly conservative imperatives into the legislation and increasingly had removed the Park Service's authority over its own property.

The third and perhaps most troubling concession was to make the Presidio Trust accountable to Congress rather than to the Department of the Interior and National Park Service. It became clear that Congress did not trust the Park Service to manage the Presidio and consequently added a provision that made the Presidio Trust's board of directors directly accountable to Congress. The provision requires the board chair, upon appointment,[97] to meet with the chair of the House committee and the chair of the Senate committee—in effect saying, "Now that you've been appointed, come and get your instructions from us, the Congress."[98] This, too, was a backhanded way of showing that Congress had little faith in or respect for the Park Service. Clearly, Congress had an agenda with regard to the national parks, and these concessions may have a long-term impact on parks other than the Presidio.

Despite these concessions, it was considered more important to get the Presidio Trust in place than to allow the bill to wither in Congress. By agreeing to these concessions, Pelosi generated the broad support needed for passage of the Presidio legislation on the House side. It appeared to many that the legislation would have the backing of Senate leadership. In fact, by the spring of 1995, House Majority Leader Newt Gingrich and Senate Majority Leader Bob Dole told the Park Service, GGNPA, and Presidio Council members that they would support passage of the Presidio Trust bill.

The Presidio legislation moved from the House to the Senate, where it faced yet another unusual obstacle.

THE PERILS OF BEING POPULAR

For several years, passage of Presidio legislation had been thwarted because the bill was perceived as unpopular. Now, the Republican House and Senate leadership was backing the Presidio legislation, and it enjoyed the broad support of Democrats; the bill was popular, and its passage was almost assured. But in becoming popular, the bill became caught in a catch-22 situation. Because politicians considered the bill likely to pass, it became a target for "rider" bills—other pieces of legislation were attached to it.[99]

The Presidio Trust legislation became a target for unpopular amendments. Senators used the opportunity to attach politically controversial legislation, such as the Utah wilderness bill, in hopes that the broad support for the Presidio legislation would carry these other more contentious amendments.

The trust legislation began as an 18-page document. By the time it reached the Senate floor, it had grown to 300 pages and included a variety of park bills and other amendments—"everything but the kitchen sink."[100] The Presidio Trust bill became enmeshed in a political struggle that had little to do with the debate about the Presidio's future. Many of these riders actually undermined its very popularity and kept the legislation deadlocked. In April 1996, Senator Ted Kennedy, D-Mass., announced he would attach his minimum wage bill to the Presidio Trust legislation. Senator Dole, now a Presidio supporter, pulled the bill off the Senate floor to prevent this from happening. Working with Senate leadership, Presidio Trust advocates reengineered the bill, removed the Utah wilderness bill, and somehow convinced Kennedy not to add the minimum wage amendment.[101]

Finally, in October 1996, after months—even years—of inching forward, the Senate approved the Presidio legislation as part of a broad package of legislation on parks and public lands known as the Omnibus Parks and Public Lands Management Act of 1996 (Public Law 104-333).[102] It was the Senate's last major act before it adjourned for the election. President Clinton signed the bill into law in November 1996. Figure 7.4 summarizes the events that had led to its passage.

In April 1997, President Clinton announced his appointees to the Presidio Trust board of directors. They were sworn in on June 25, 1997, in a special ceremony in San Francisco. Toby Rosenblatt, long-time advocate for the Presidio and chair of GGNPA, was appointed president of the board of directors. Other appointees included Dr. Edward Blakeley (Dean, School of Urban and Regional Planning at the University of Southern California and former member of the Presidio Council); Donald Fisher (Chair, Gap, Inc.); Amy Meyer (Chair, People for a Golden Gate National Recreation Area); May Murphy (Partner, Farella, Braun and Martel—specializing in real estate law); William K. Reilly (former EPA administrator and former senior adviser to the Presidio Council); and deputy Interior Department secretary John Garamendi (the designee of Secretary of the Interior Bruce Babbitt).[103]

Figure 7.4
Timeline of the Presidio Trust Legislation

1993

Summer	Department of Defense decides to keep small contingent of Sixth Army at Presidio.
	Representative Duncan proposes amendment to reduce Presidio budget for 1994. Amendment defeated, 230 to 193.
Oct.	National Park Service unveils Presidio plan; begins 60-day review period.
	Representative Pelosi announces bill to create Presidio public-private partnership to be called "Presidio Corporation."
	Initial House subcommittee on parks hearing for Presidio legislation.
Nov.	Pelosi formally introduces legislation, H.R. 3433, to House.
	Congress approves Defense Department decision to keep Sixth Army at Presidio; Army informs Park Service of intention to keep the golf course.

1994

Jan.	Presidio Council and Department of the Interior recommend name change to Presidio Trust.
April	National Parks and Conservation Association sponsors an Earth Day "March for Parks."
	San Francisco Board of Supervisors passes resolution supporting H.R. 3433 and another resolution against Duncan's legislation, H.R. 4078.
	GGNRA Advisory Commission approves the Park Service's Presidio plan.
May	Congressional hearings on H.R. 3433 (Presidio Trust) and H.R. 4078 (Duncan bill).
	Duncan reintroduces amendment to reduce Presidio funds for 1995. Amendment defeated, 257 to 171.
June	House Energy and Natural Resources Committee approves H.R. 3433 (Presidio Trust) by 28 to 14. Bill moves to full House for a vote.
Aug.	August 18, House of Representatives votes to approve H.R. 3433. Bill moves to the Senate.
Oct.	October 1, official transfer from post to park.
Dec.	Base Realignment and Closure Commission recommends the Pentagon eliminate Sixth Army entirely for further defense savings. No Army troops will remain at Presidio.
	University of California at San Francisco pulls out of Letterman Complex negotiations with Park Service. Presidio loses prospective anchor tenant.

Figure 7.4 Continued

Dec.	Senate filibuster of Presidio Trust legislation. 103rd Congress concludes second session and adjourns. Bill "killed."
1995	
Jan.	104th Congress convenes; Pelosi reintroduces Presidio Trust bill, now H.R. 1296/S. 594.
Sept.	House of Representatives approves Presidio Trust. Bill moves to Senate.
Dec.	First session of 104th Congress concludes, and Presidio Trust legislation is not acted upon.
1996	
Oct.	Presidio Trust legislation, with Senate and House leadership backing, finally passes both houses as part of a larger national parks legislative package. President Clinton signs bill into law in November.

The Politics of Parks

Realizing the Grand Vision entailed a complex political struggle at both the local and the national level. At the heart of the struggle to pass the Presidio Trust legislation and to defeat Duncan's bill was a broad, fundamental debate about the meaning of parks in American society. The debate revolved around expectations of what a park is, what it should be, and what the responsibility of the National Park Service should be in managing a park. The politics of the Presidio was also a test case for a political philosophy that sees all national parks as a resource to exploit for money. Duncan's overt political challenge can be seen as part of an ideological movement that seeks divestiture of national parks and public lands for economic reasons.

The political debate about the Presidio legislation marks the first time a national park was asked to pay for itself or face disposal. It also raises the question: what features make expensive national parks acceptable? (Table 7.3 selects a sample of national parks for comparison.)

Is it merely size? Yosemite and Yellowstone, the two most expensive parks in the system, are thousands of acres of wilderness and hundreds of times larger than the Presidio. If size were the only measure of worthiness, then resistance to the Presidio would seem logical.

Is it visitor/public access? The Presidio currently attracts more visitors each year than either Yosemite or Yellowstone. In fact, the GGNRA ranks

Table 7.3
Measuring Merit?

Park	Visitation (in thousands)	Size (acreage)	Annual Budget
Yellowstone	2,934	2,219,791	$17,404,000
Yosemite	3,809	761,236	15,430,000
Grand Canyon	4,530	1,217,158	11,214,000
Everglades	898	1,506,499	8,102,000
Blue Ridge Parkway	17,910	88,159	8,653,000
Mt. Rainier	1,396	235,613	6,660,000
Gateway NRA*	2,281	26,579	13,615,000
Golden Gate NRA	**16,723**	**73,180**	**9,481,000**
Chattahoochee River NRA	2,841	9,260	1,519,000
National Capital Parks	14,289	7,310	23,179,000
Statue of Liberty	4,110	58	8,322,000
Independence Hall	3,141	45	9,015,000

Source: Compiled by author from U.S. Department of the Interior, *Budget Justification, F.Y. 1995, National Park Service* (Washington, D.C.: National Park Service, 1994: 160–172).

Note: The parks in this table represent the top five park units for either visitation, size, or budget.

*NRA stands for "National Recreation Area."

second in visitors per year with over 20 million; Yosemite and Yellowstone rank fifteenth and sixteenth respectively with approximately 5 million visitors each. The Presidio's urban location promises to provide recreation opportunities for an entire metropolitan region; Park Service officials estimate that by 2010, more than 10 million people will visit the Presidio each year.

Is it capacity for use? Arguably there are as many multiuse activities at the Presidio as at the large wilderness parks. These include traditional park activities such as walking, hiking, biking, fishing, and picnicking, and nontraditional activities such as boardsailing, golfing, tennis, attending conferences, job training programs, and education at the global center.

Is it capacity for education? The proposed global environmental education center marks the first effort by a national park to expand its leadership in environmental education.

The struggle to realize the Presidio-as-park is part of a national debate about the purpose and importance of parks in urban America and the responsibilities of the National Park Service. This is the focus of discussion in the next chapter.

CHAPTER 8

Nature, Culture, and the National Parks

In January 1996, the Golden Gate National Park Association (GGNPA) launched a new campaign to educate the Bay Area community about improving and preserving the parklands of the Golden Gate National Recreation Area (GGNRA). The campaign had three goals: first, to increase public awareness of the obstacles facing parks, like cutbacks in government funding; second, to encourage a public dialogue about the parklands themselves; and third and most important, to create, through association membership and donations of time and money, a sense of community and personal ownership of the GGNRA parklands. GGNPA hoped the campaign would change the widespread perception that the Golden Gate National Recreation Area was understood and enjoyed only as a collection of individual sites rather than as one large entity.

A highlight of the campaign was GGNPA's introduction of a new name for the GGNRA: the Golden Gate National Parks. The GGNPA newsletter explained the name change: "We believe this name is not only descriptive of the parklands of the Golden Gate but is easy to refer to and remember. *It also clearly acknowledges this park's status as a unit of the national park system, clarifying this issue for many who are confused by the term 'recreation area'*" (my emphasis).[1] Responding to this name change, local graphic artist Michael Schwab (who had created the design for the Presidio's Post to Park logo) designed a new "master logo" to serve as the central unifying image for the new campaign (see Figure 8.1). "The new theme and accompanying graphic image recognize the beauty of its individual places and the importance of the whole."[2] In addition to the campaign, other changes ensued. The Golden Gate Na-

Figure 8.1
The Golden Gate National Parks Logo. Michael Schwab, who designed the Presidio's Post to Park logo (see Figure 7.3), was asked to design a new "master logo" to promote the parks. It is interesting to note that the National Park Service has never developed a graphic campaign for any of its parks. Schwab also designed images of the Presidio, Alcatraz, Marin Headlands, Muir Woods, Fort Point, and Lands End (the Cliff House), to show the diversity of the parks. *Source: The Golden Gate National Parks Association.*

tional Park Association added an "s" to "park" in its name; the organization is now called the Golden Gate National Parks Association. And the quarterly GGNPA newsletter, previously titled *The Park,* was rechristened *Gateways.*

The GGNPA campaign would involve a series of public service announcements to air on local television stations, as well as ads and promotional activities at the various park visitors centers. GGNPA hired the San Francisco–based advertising agency Goodby, Silverstein & Partners, perhaps best known for its "Got Milk" ads. Rich Silverstein, co-chair of the firm (and GGNPA board of trustees member), described its creative approach: "our goal is to use humor to get people's attention, then motivate them to invest—on a personal level—in the Bay Area's national parks."[3]

The campaign, the name changes, and the new logo tell us much about the current state of the national park system, the Golden Gate National Recreation Area, and, by extension, the Presidio. It is at once symbolic and tangible. According to surveys done by GGNPA, the name change was needed because many Bay Area residents did not perceive the Golden Gate National Recreation Area as an integral part of the national park system. This confusion was reinforced by the political struggle in Congress over the legislation, a struggle that appeared to question the "worthiness" of the Presidio to be part of the national park system and indeed the integrity of the entire GGNRA.

Shaping the National Park System

The history of national parks has attracted the interest of many scholars and park enthusiasts, including Horace Albright, Alston Chase, Lary Dilsaver, Ronald Foresta, George Hartzog, Jr., Hal Rothman, Alfred Runte, Douglas Wellman, and Conrad Wirth.[4] These writers have situated national parks in a broad social context and have explored the parks themselves, the contributions of park creators and administrators, and even the very terms *park* and *wilderness*. They have shown the ways in which Americans' attitudes toward the outdoor environment have changed and how these attitudes have influenced both the creation of parks and park ideology. For our purposes, it is necessary to outline the history of national parks in order to place the Presidio struggle in a larger political and social context.

Let me begin with a generally accepted position: there has been no universally agreed-upon definition of what constitutes a national park—it is not explicitly defined in the original legislation, nor has it been clarified in more recent efforts to reconsider the mission of national parks. Rather, this has been a philosophical issue debated throughout the history of the parks, and it remains a highly contentious subject. Currently, the national park system comprises some 370 units of more than 80 million acres. These 370 units fall into approximately sixteen different categories of parks (see Table 8.1). The quantity and diversity of the national parks is the outcome of more than a century of expansion and evolution within the national park system. Yet despite the evolution of the national park system, little information exists that defines its nature and substance.[5]

Many processes have been shaping the national parks. This chapter focuses on two that have helped to define and redefine the purpose, function, and ideology of national parks, and have had a relevant influence on the current state of urban national recreation areas, such as the GGNRA.

Table 8.1
Units of the National Park System

Number	Category
51	National Parks
23	National Battlefields / National Military Parks
4	National Battlefield Parks / National Battlefield Sites
32	National Historical Parks
71	National Historical Sites
76	National Monuments
26	National Memorials
13	National Preserves
9	National Wild and Scenic Rivers and Riverways
7	National Rivers
18	National Recreation Areas
14	National Seashores and National Lakeshores
3	National Scenic Trails
1	International Historical Site
4	National Parkways
11	"Others" (includes National Capital Parks, the White House)
363	**Total**

Source: Compiled by author from National Park Service, *National Parks for the 21st Century: The Vail Agenda* (Post Mills, Vt.: Chelsea Green Publishing Co. / National Park Service, 1993: 9); and National Park Service, Office of Public Affairs and Division of Publications, *The National Parks: Index 1989* (Washington, D.C.: GPO, 1990).

Note: A tabulation of the park units and park designations reveals that there are more "culture" parks than "nature" parks, an interesting and perhaps surprising fact that is discussed later in the chapter.

The first process is the periodic emergence of movements to reconsider the criteria by which parks receive the designation "national park." These movements have introduced federal legislation altering the criteria and thus leading to the creation of new types of national parks. Several of these movements have been efforts to shape a meaningful system, to modify the structure of the national parks, and to articulate how Americans view the national parks.

The second process is the eruption of political conflict over specific proposed or existing parks. Often struggles have revolved around issues of development rights or management policy, but equally as often they have fundamentally challenged the national park ideal and purpose. These conflicts, both local and national in scope, have arisen as a result of the publication of

Park Service documents, studies, or ecological surveys, and the initiative of individuals or organizations such as congressional representatives, intellectuals, park supporters, park opponents, and nonprofit environmental organizations (such as the Sierra Club). Indeed, as the legislation put together by Congressman Philip Burton indicates, one person can make a difference for the national parks. Political struggles and public debates too have shaped and redefined the National Park Service and perceptions of national park ideals. Such debates took place in the face of changes in public and government attitudes and values about a variety of issues, including economic development, preservation, protection, and recreation.

Together these two concurrent processes—park movements and park political struggles—have influenced the meaning of parks. An understanding of these two interdependent historical forces situates the current struggle for the Presidio in the broader context of national park evolution and change.

Park Movements and the Designation of New Parks

Historically, four broad movements have shaped the contemporary national park system by creating laws that established new parks and set forth policy.[6] Each movement arose for a different reason, mirroring the political, intellectual, cultural, and geographical imperatives of the time. Each defined a new type of national park unit, expanding the national park system and increasing the ways in which people use, visit, and think of national parks and conceive of the national park system.

The first movement, known as the conservation movement, began in the late nineteenth century. It has been well documented and described by authors such as Hans Huth, Roderick Nash, and Joseph Petulla.[7] Out of this movement the first national parks, Yellowstone in 1872 and Yosemite in 1891, were created.[8] The origins of how and why the movement began are shrouded in uncertainty. It is probable that the movement to create the first national parks arose as a complex aggregation of several cultural forces, including responses to political, economic, and geographical realities of the nineteenth century.

First, the influence of the transcendentalist movement helped to shape the attitudes of educated, upper-class Americans during the late nineteenth century.[9] Second, the influence of naturalists such as George Perkins Marsh increased understanding of natural systems and human impact on them. Third, the rapid settlement of the western frontier, along with increasing industrialization and urbanization in the East, prompted many intellectuals to reevaluate their attitudes about "nature" and "culture."[10] While men and women battled the wilderness during the late nineteenth century, a romantic movement was conceived, ironically, in the cities.[11] This movement promoted an appreciation for the beauty in nature and saw the natural world as "the handiwork of God if not His very image."[12] Although many Americans continued to see the wilderness as worthless or something to be transformed, the writings of romantic transcendentalists influenced others to see it as something aesthetically valuable, a spiritual asset, and a source of American pride.

Gradually, the conservation movement began to grow beyond the shadows of a few naturalists, philosophers, and writers. Eventually, the movement coalesced into an effective political coalition and called on Congress to set aside vast areas to be designated national parks. In March 1872, President Ulysses S. Grant signed the bill creating the first official national park, Yellowstone. The designation of Yellowstone and Yosemite as the first two national parks thus defined the ideal qualities of a national park.

While intellectuals were looking at nature and wilderness from a new perspective, America was searching for a national identity. The historian Alfred Runte chronicles the meaning of national parks in American history in his book *National Parks: The American Experience*. His central argument is provocative: the driving force behind the movement to establish national parks arose not out of environmentalist or conservationist ideology as much as from a reaction to the dearth of recognized cultural achievements in the young nation in the nineteenth century. Unlike Europe, America had no castles, no ancient ruins, no Shakespeare, and no Sistine Chapel.[13] To compensate, Americans turned to monumental scenic landmarks as substitutes for human achievements and as America's contribution to world culture. Great chasms, thundering waterfalls, and towering peaks became America's secular cathedrals and were considered national assets. These natural monuments and the creation of national parks were considered uniquely American. They were seen as contributions to humanity in the same manner as the Swiss Alps, Stonehenge, or the Parthenon.[14] Runte concludes that the birth of the national park movement was inspired by monumentalism. Monumentalism in mountains, canyons, gorges, and other dramatic geologic and topographical features thus became the predominant national park ideal. Monumentalism was in fact linked to quasi-aesthetic romantic notions of the sublime.[15] Dramatic scenery inspired deep emotion and became ingrained in the American imagination. Places such as Yellowstone and Yosemite became the archetypes of "nature" and "wilderness."

Runte's argument is important and intriguing, for it means that both the ideals behind the national parks and the national parks themselves are a *cultural expression*. National parks are social constructs—embodiments of the values a society deems important, time capsules that reveal attitudes and perceptions dominant at the time of their creation.[16]

It is also important to note that the creation of any national park is somewhat arbitrary and artificial. The delineation of park boundaries, for example, has often been contested, debated, and compromised. Especially in the early years of the parks, boundaries enclosed a space defined as a national park although the land surrounding these boundaries was, in many cases, identical to the land within (Yellowstone and Yosemite are examples).[17] Thus park boundaries are the product of a social process. In this regard, there is much culture and politics in places that many consider "nature."

National parks, people, and nature exist in an intricate arrangement of political, social, legal, intellectual, and sentimental relationships.[18] Parks are as much about statesmanship, philanthropy, and cultural values as they are about ecology. This was true of the origins and founding ideology of the first national parks. It is also true of each and every subsequent movement to expand and modify the park system.

From their inception, national parks have been enmeshed in debates about the accepted definition of parks. Should they just be isolated wilderness, or should they include other types of places significant to the American identity? The division embedded a tension between purpose and ideal in the park from the earliest years.

On one issue there was little debate: national parks, from the beginning, were not about whether to use or not use the land. Park advocates did not intend the parks to remain idle. Rather, use of national parks was fundamental to the national park ideal: public access and enjoyment would illustrate (and legitimize) the park's worth to the nation. The concept of use evolved to include activities such as inspiration, education, and recreation. From the beginning, then, public use was an important element of park philosophy. Secretary of the Interior Hubert Work reinforced this ideal when he wrote in 1925 that "the public should be afforded every opportunity to enjoy national parks and monuments. . . . parks and monuments should be kept accessible by any means practicable."[19] While ideals of pristine wilderness, preservation, and dramatic scenery remain ingrained in our twentieth-century imagination as founding park ideals, it is important to remember that public access, recreation, and use are also such ideals.

Nature Versus Culture: The First Debate

As the citizens of the United States began to reconstruct the nation out of the ashes of the Civil War, the second park movement began as a postwar project sponsored by political elites. Amid the political and social uncertainty of Reconstruction, political elites recognized the importance of doing something to heal the nation's war wounds and to commemorate all those who fought in the Great Rebellion. The Civil War was seen as a defining moment in the American story, and many believed that battle sites should be preserved. Thus this park movement provided the earliest efforts to preserve "sacred ground."[20] In 1890, only a few years after the creation of Yellowstone, Congress passed the National Military Parks and Battlefield Act establishing two national battlefields, Chickamauga (in Georgia) and Chattanooga (in Tennessee). These places were to stand as lasting memorials to the great armies of the Civil War. They might also be considered the second and third national parks (although at the time they were not called parks nor were they considered part of the informal park system).

The inclusion of battlefield sites and military parks as national parks was a departure from the monumentalism ideal. Rather, the leaders of this movement aimed to designate and protect sacred ground and to set aside certain places as profoundly significant in American history. The valuing of cultural significance meant that natural features were not necessarily the overriding factor determining whether a space was an important part of American history and identity. Although the romantic outlook never disappeared completely, this park movement introduced notions of historical significance in identifying places worthy of park status. It was important in articulating an emerging tension: the conflict over whether cultural areas measured up to the national park ideal. This tension between parks as nature and parks as American history and culture would prove to be important and enduring.

The Role of Culture in the Parks Expands

The new century brought accelerated changes to national parks. In 1906, Congress passed the National Monument and Antiquities Act, an achievement that arose out of the third movement to expand and modify the meaning of national parks. This legislation was designed to more clearly articulate and protect American history by including both natural and man-made areas of significance.[21] In part the act was a response to growing concern about the desecration of Native American ruins in the Southwest. The National Monument and Antiquities Act expanded preservation to include historic landmarks, historic and prehistoric structures, and archaeological sites (notably prehistoric Native American sites). For the first time, Americans recognized pre-Columbian cultures as an important part of American history.

The interest in preserving and gathering objects or sites of historical significance or antiquity was an important driving force behind the legislation. Among the new sites designated as national memorials or national historic parks were Chaco Culture National Historic Park in New Mexico (1906); Gila Cliff Dwellings in New Mexico (1907); Muir Woods in California (1908); Oregon Caves in Oregon (1909); Devils Towers in Wyoming (1906); Salinas Pueblo Missions in New Mexico (1909); and Abraham Lincoln Birthplace in Kentucky (1916). The National Monument and Antiquities Act was one of the most important pieces of preservation legislation enacted by the U.S. government. Without it, there would have been little flexibility in the preservation process, and many areas of significance would have been destroyed before Congress passed legislation to protect them.[22]

The Battle Sites and Military Parks Act, and the National Monument and Antiquities Act testify to the increasing interest in preserving, protecting, and showcasing American culture. The passage of this legislation effectively reorganized national parks by including new types of units. Despite the still predominant view that national parks were natural monuments and scenic wonders, significant actions had been taken to broaden ideas of appropriate places and consider culturally historic sites.

Defining a Park System

During the first decades of the twentieth century, the number of national parks increased, and the criteria for designation as a national park broadened. But despite the broadening of criteria, the national park ideal remained rugged, spectacular landforms.[23] Romantic notions of nature and wilderness as sublime, and the ideology of monumentalism, remained the preeminent forces shaping American perceptions of national parks, even though recent additions (such as the Lincoln Birthplace and the Aztec Ruins in New Mexico) reinforced a distinctly cultural definition of parks.

Regardless of the criteria for the newest sites, all parks were without an administrative structure to both manage the parks and provide advocacy. Financial support and staff were insufficient. Congress had created and named several places as national parks but had not established an institutional framework to manage and administer the parks; there was no park system per se. Each unit was separately administered, depending on its location and purpose, by the secretary of the interior (the "nature" areas of Yellowstone, Yosemite, and others), the secretary of war (historic battlefields and military parks), or the secretary of agriculture (historic sites and monuments).[24]

Many park advocates believed that the lack of an institutional administrator left the parks vulnerable to political whim. Park supporters worried that politicians were too quick to yield to economic development projects by either deactivating areas of parks or allowing commercial activities on park land. Finally, in 1915, Congress established the National Park Service, and President Woodrow Wilson signed the National Park Service Act into law, bringing more than thirty-six national parks into the park system. The legislation was the culmination of long-standing attempts to integrate and incorporate the various units already designated as national parks. The act was also seen as the best hope of protecting the parks against the uncertainties of the political climate.[25] This was no idle threat; since their inception, the parks had been under constant pressure from development interests. The legislation, however, attempted to clarify the role of the new institution, charging the National Park Service

> to promote and regulate the use of the Federal areas known as national parks, monuments and reservations hereinafter specified by such means and measures as conform to the fundamental purpose of the said parks, monuments and reservations, which purpose is to conserve the scenery and the natural and historic objects and the wild life therein and to provide for the enjoyment of the same in such manner and by such means as will leave them unimpaired for the enjoyment of future generations. (National Park Service Act, title 16, sec. 1)

The legislation creating the National Park Service contained a glaring omission. The culture sites were not included as parks under the purview of the National Park Service. The new system contained only nature parks. Yellowstone, Yosemite, and other national parks finally belonged to a system. But at no point could anyone state with any sense of final authority that "this is what a park should be."[26]

Continued Expansion of Both Nature and Culture Parks

Franklin Delano Roosevelt's New Deal intricately linked socioeconomic programs with Park Service programs. During much of the 1930s and 1940s, despite the Great Depression and the nation's economic troubles, the Park Service expanded its agenda. New Deal programs such as the Civilian Conservation Corps profoundly impacted the parks through public works projects such as road and trail construction and the building of hotels and other park accommodations.[27] The Works Progress Administration also engaged in infrastructure repair and construction at many parks across the country. Thus the programs of the New Deal improved the physical condition of parks and provided impetus for park expansion. Along with physical repair projects, there was renewed attention to places of historical significance.

Roosevelt's New Deal programs did much to improve the condition of many of the natural parks, but perhaps his most significant action occurred in 1933 when he transferred national monuments, national battlefields, and national cemeteries into National Park Service jurisdiction. The move was prompted by the National Park Service director, Horace Albright, who saw that the inclusion of these cultural assets would make the Park Service truly a national agency and would give it a broadened constituency that would ensure more attention (and money) from Congress.[28] Roosevelt's reorganization thus enlarged the domain of the national park system by fifty-six parks. However, many Park Service professionals were not happy about the inclusions.

After this reorganization and because Americans were becoming more aware of and confident in the emerging narrative of the American story, Congress passed the Preservation of Historic Sites Act in 1935 to recognize the wealth and diversity of places and people who had contributed to the American identity. The legislation broadened the Park Service's sphere of influence in historic preservation. It incorporated places such as Mount Vernon and Monticello into the national park system, declaring these sites historically important and therefore nationally significant. Although many new parks were being created because of their scenic monumentalism, parks had also become the time capsules of American culture.

In the decades following the Second World War, the Park Service witnessed a rapid expansion of properties and a doubling, even tripling, of visitors to the parks. While the park system was expanding, it was also responding to the political, economic, social, and technological changes in American society. The postwar years were ones of economic prosperity and were characterized by industrial expansion and urban/suburban development. The increasing affordability of automobiles and the concomitant mobility changed the way in which people spent their leisure and also stimulated new attitudes about the out-of-doors. The increase in numbers of visitors to the parks raised concerns about access.

By the 1950s, many park advocates and even Park Service administrators felt the park system had grown beyond its ability to provide adequate staff, facilities, and protection. As a result of growing concerns about the physical ability of parks to cope with so many visitors, the Park Service launched "Mission 66," a program designed to upgrade the condition of parks through a series of projects such as increasing staff, building and repairing roads, and investing federal funds in scientific surveys and studies.[29] Although the Park Service recognized the inadequacies of many of the parks, it felt pressure to continue expanding programs and adding new parks. This pressure came from political elites who realized the political benefits they would derive from creating a national park for their constituencies.

Parks for People: The National Recreation Areas

The fourth movement to expand and modify the park system was, in part, a result of executive-branch and congressional concerns about recreation and access to national parks. In the late 1950s, Congress had authorized an Outdoor Recreation Resources Review Commission to study the problem of open space and recreational opportunities for the nation. Congress was responding to the increasing sprawl of urban centers, highways, and residential, commercial, and industrial development around the nation. In 1962, after three years of research, the commission issued its report, which stated that outdoor opportunities were *most urgently needed near metropolitan areas* (my emphasis).[30] Further, the commission found that simple activities—walking, hiking, and picnicking—were the outdoor activities in which Americans most participated. The report concluded that few recreation areas were near enough to metropolitan areas for a Sunday outing and said, "the problem is not one of total acres, but of *effective* acres."[31] The report noted that the federal government was in a good financial and leadership position to implement the creation and establishment of recreation areas around the country and recommended that the National Park Service begin planning for national recreation program units. These national recreation areas would be located in or near urban areas and designated primarily for outdoor recreation use rather than natural or historic preservation. This recommendation would increase National Park Service responsibility and suggested new criteria for parks designed expressly to accommodate high use.

The fourth and most recent national park additions thus were national recreation areas—seashores, lakeshores, and urban recreation areas. This particular park movement was a response to the perceived problem that "wilderness" in Park Service terms was neither available nor accessible to most of the urban population (see Table 8.2). Thus this movement reveals the political influence of major metropolitan areas and the eagerness of the politically ambitious to harness this influence, the increasing concern for open space and environmental quality, and the increasing acceptability of federal support in matters of local concern.[32]

The first recreation-oriented park to be created was the Cape Cod National Seashore. With the authorization of this new unit, Congress took a significant pioneering step for the national park system. The establishment of the Cape Cod National Seashore set an important precedent for the creation of what would be referred to as nontraditional parks—seashores, lakeshores, and urban recreation areas. Both policy and ideological issues emerged from congressional debate about the creation of a national seashore. First the Park Service was challenged to learn new ways to accommodate the concerns of nearby communities, something it had not had to do very often in the more remote and isolated parks. Many of the national lakeshores, seashores, and

Table 8.2
Public Recreation Areas, 1965 (in millions of acres)

	Urban	Nonurban	Total
Federal	35.9	410.7	446.6
State	4.3	35.4	39.7
County	.7	2.3	3.0
Municipal	1.4	.6	2.0
Total	42.3	449.0	491.3
Percent	9%	91%	100%
Population*	123,813,000	68,372,000	192,185,000
Percent	64.4%	35.5%	100%

Source: Conservation Foundation, *National Parks for the Future* (Washington, D.C.: Conservation Foundation, 1972: 76–77).

*Population estimate based on Bureau of the Census, "Population Report," 1966.

urban recreation areas were originally private lands that were either donated to or purchased by the Park Service.[33] Second, and perhaps most important, the creation of the Cape Cod National Seashore, located within a day's drive for nearly one-third of the American population, sparked a major philosophical debate about national park criteria.

Not long after the creation of the Cape Cod National Seashore, other seashores and lakeshores were designated. In the early 1970s, under President Nixon's "Parks to the People" program, several urban recreation areas were established.[34] On the day Congress authorized the Golden Gate National Recreation Area in and around San Francisco, it also authorized the Gateway National Recreation Area in the New York and New Jersey metropolitan area. The outreach of the national park system broadened with the addition of these two urban national recreation areas (NRAs). Urban NRAs were significant in several ways. First, they reflected new criteria. Second, part of establishing and justifying the new criteria required a redefinition of *national significance* to include areas not only unspoiled by development but also accessible to the public. Thus the establishment of subsequent recreation areas, like the GGNRA, was justified on the basis of public access, historical significance, and natural preservation. The urban NRAs were also significant because prior to their creation nearly all the Park Service's holdings in major urban areas (outside the national capital region) had been small historic sites, where the primary concerns were historic preservation and interpretation. The Gateways challenged the Park Service to manage urban mass recreation, a task not previously a federal responsibility.

The NRAs stirred debate both inside and outside the Park Service. Some felt they would undermine national park values and jeopardize parks of the traditional sort (nature parks).[35] Others felt they were deserving and believed it was necessary to secure equal protection and equal appreciation for all units of the system. Underlying the debate was a vague perception that the concepts and ideals had strayed from the original park ideology. No longer did a place need to be monumental or even green to be included. Opponents of the national recreation areas argued that the criteria for selection were different from (and thus inferior to) those used to designate traditional parks, because man-made features such as parkways, reservoirs, and parklands were included. Such features were not sublime.[36] In the opinion of many Americans (and members of Congress) commonplace topography did not inspire nature worship and thus did not belong in the park system.

Opponents argued that urban NRAs possessed only local, not national significance. Previously, the issue of scale (local or national value) had not been debated with any vigor. With the addition of urban NRAs, however, conflict about geographical scale and about what constituted national or local significance ensued.[37] Opponents questioned whether a city could have a park of national significance. Recreation areas thus challenged the national park ideal and purpose by challenging the definition of *significant* as well as *national*. The concept of national significance has been a guiding principle for national parks, but it is an ambiguous concept that has changed throughout park history. Just as the creation of any national park is a reflection of social and political influences at a given point in time, so too is the concept of national significance. National significance is in the eye of the beholder.[38] To many, it seemed that urban spaces were unlikely to possess attributes of national significance. However, judging a park's merit from its location is problematic. This difficulty has been wonderfully articulated by Dwight Rettie:

> Yosemite is an acknowledged nationally significant area. . . . well over half of Yosemite's visitors are Californians, and a majority of those come from the San Francisco metropolitan areas. Does that fact make Yosemite merely a regionally significant site? Golden Gate NRA, located entirely within the San Francisco metropolitan area, contains attractions that appeal primarily to people from outside the region—Alcatraz Island, for example, whose visitation is estimated to be over 90 percent non-Californians.[39]

The inclusion of recreation areas clearly demonstrated a change in the national park ideal. For many years, the overriding criteria for designation as a national park was the presence of natural wonders—proposals for new parks were weighted on the basis of physical endowments—or the occurrence of events important in American history. Recreation areas satisfied neither criterion exclusively yet often had elements of each. *They were a new category: neither nature parks nor culture parks.* They floated ambiguously outside the national park ideals. Opponents feared the national park system would become a collection of mediocre places and insisted that urban recreation areas would strain the Park Service budget and thus dilute the agency's effectiveness in managing other parks.

Debate over the inclusion of national recreation areas was one of the first instances in which nonprofit environmental organizations opposed rather than supported the designation of new parks. One of the most vocal opponents was the nonprofit environmental organization the Conservation Foundation, which voiced concerns about recreation areas in its 1972 report on national parks.[40] Its task force stated that it believed national park programs had lost their focus and had responded to too broad an array of stimuli: "They have tended to fill vacuums, and they desperately need to rediscover a unifying ethic."[41] The report suggested that efforts to make the Park Service more inclusive had only diluted the mission of environmental preservation. The report recommended that parks designated as national recreation areas be eliminated from the jurisdiction of the Park Service because they deemphasized preservation and focused too much on recreation, and it concluded that urban recreation areas should be the responsibility of the federal government, but not the Park Service. "Many existing units of the National Park System do not fit what we believe the system ought to be in the future," noted the report.[42] The Conservation Foundation considered the Golden Gate and the Gateway national recreation areas "anomalies" and, while noting that these areas possessed parklike values, observed that

> nevertheless these projects, as conceived, are not *intrinsically* National Parks and would require services and facilities which are quite different from those found in traditional resource-based park units. *We recommend that the Gateways be transferred as soon as possible to appropriate state or regional agencies for administration* [my emphasis].[43]

The Foundation's report suggested limiting the national park agenda to exclude urban areas.

On the one hand, the Foundation was justified in its concern that the Park Service agenda had broadened beyond the agency's ability to provide efficient and effective management. On the other hand, the Foundation offered a narrow and somewhat elitist view of what a park should be. After all, Thoreau's Walden was relatively close to an urbanized, industrialized part of America, and Thoreau valued Walden not for its acreage (the pond was in the midst of about 61 acres and was only about 1.5 miles in circumference) but as a setting or environment that transported him into a different way of being.

The position of the Conservation Foundation and the recommendations of its report highlight the conflicts within and outside the Park Service over national park ideals. The Foundation clearly believed that national recreation areas were not worthy of national park status. This belief, in my judgment, foreshadowed decades of second-class status for the NRAs.

The Crown Jewel Syndrome

Despite more than a century of evolution and additions to the park system, Yellowstone and Yosemite continue to epitomize the national park ideal.

Dwight Rettie discusses the crown jewel syndrome and notes that the label "crown jewels" is reserved for a select number of national parks—Yellowstone, Yosemite, Grand Canyon—places that have a special mystique.[44] Not surprisingly, these are nature parks. Indeed, the accolade "crown jewel" would not be applied to even the most renowned and revered historic sites, such as the Statue of Liberty or Independence Hall.

The crown jewel distinction has several repercussions. First, it establishes a national park hierarchy. At the top are the so-called crown jewels; below them are all the rest. Second, the distinction implies, and often is used to suggest, that some national parks have a better claim on the Park Service budget than others.[45] Third, casting parks in a hierarchy makes it easier to disregard the lowest-ranked parks. In times past, various national park supporters have advocated that the national park system be limited to the crown jewels and that all other units in the system be transferred to another agency.[46] When Park Service professionals themselves express doubt about the worthiness of a park's existence or about its cost, this goes beyond mere hierarchical thinking. It jeopardizes the park system's diversity. In addition, the crown jewel syndrome may result in disproportionate allocations of personnel, money, and sentiment.

The concept of a hierarchy within the park system, however, is problematic. A hierarchy implies a well-defined set of categories, but the actual makeup of a park hierarchy has never been described, perhaps because not all Park Service professionals subscribe to the notion of a park hierarchy. Nor has such a hierarchy ever been articulated as official park policy. It is likely, then, that a national park hierarchy can take a variety of forms. I present one possibility in Figure 8.2.

The hierarchy depicted in Figure 8.2 might best be described as a quasi-trichotomy.[47] It shows two dominant categories of parks: nature and culture. Many parks meet natural-area criteria, and other parks meet cultural definitions. *Recreation areas, however, are neither yet both.* They remain outside either of the two park ideals. Herein lies the complexity and contradictory nature of urban recreation areas. The GGNRA and the Presidio, for example, contain both natural and cultural features; yet their designation reflects recreational criteria. The Presidio is not merely a cultural site containing a wealth of military history; it also contains several rare and endangered plant species, spectacular ocean vistas, and meadows for quiet contemplation. It is simultaneously nature, culture, and recreation. It is, Park Service director Roger Kennedy eloquently remarked in 1993, "a community within a park within a larger community, forming concentric circles of relationships to others and to the natural world."[48] Thus it must be placed not in rigid categories or

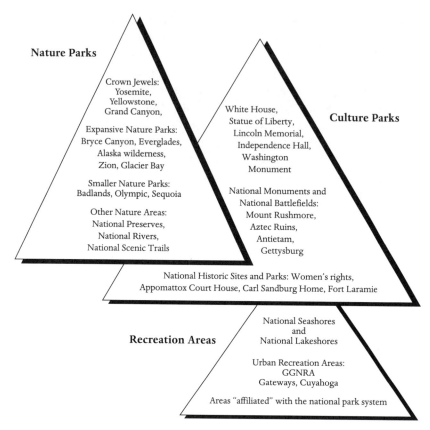

Figure 8.2
A Hierarchy of National Parks. This illustration shows one of the many possible ways the hierarchy might be formed. Note that the nature parks are set highest, indicating their predominance in the American imagination. This hierarchy conveys the categorized and often ambiguous relationship among the various parks. It also presupposes a "bottom rung" in each category. The placement of recreation areas is problematic—it is precisely their ambiguous status as neither exclusively nature nor exclusively culture that has, I argue, led to their marginal position in the park system. This illustration is not meant to be static or fixed; with the exception of the "crown jewels," a particular park's position in the hierarchy might depend on changing attitudes in the National Park Service or an individual's own perception of the national park ideal. *Source: Diagram by author.*

boundaries but in fluid concentricity, a concentricity that does not allow the neat distinctions within the national park dualisms or hierarchy.

Shown the hierarchy depicted in Figure 8.2, many Americans probably would rank Yellowstone and Yosemite as more significant than Appomattox Court House or the Aztec Ruins. It is also probable that many would rank a renowned historic site such as the Washington Monument (culture park) as more significant or more important to the park system than Hot Springs, Arkansas, a nature park. However, few would list any culture park as more significant than the crown jewels or many of the second-tier nature parks. Thus the division between nature and culture parks is not a straightforward dichotomy but has its own complexities and ambiguities. The third category in the hierarchy, recreation areas (urban, seashore, or lakeshore), presents a more problematic challenge. Given the history of acrimony and contention, it is likely that many would situate the recreation areas below many of the nature and culture parks. Since recreation areas are not explicitly nature parks or culture parks, they remain on the margins, characterized by an ambiguity of status, and they face outright hostility and contempt from both inside and outside the Park Service.

Of course, any interpretation of a hierarchy depends on one's knowledge of national parks, perspective on the national park ideal, and interest. For example, an avid Civil War historian would undoubtedly rank national military parks and battlefields as significant, but a naturalist might consider the expansive wilderness areas to be of greater significance.

Although the concept of hierarchy remains both unofficial and difficult to confirm, it remains a potent and effective subtext that can be employed by Park Service administrators and others (like Congress) as a measure of opposition to the units that are not considered to be crown jewels.

Opponents have labeled recreational areas "playgrounds," implying these parks are not historically, culturally, or naturally significant.[49] Images of physical grandeur and remote wilderness still resonate deeply within American culture, and so it is not surprising that many still devalue recreation areas. America's continued preoccupation with scenic monumentalism has obscured the value of other park units, including historical sites but most notably urban recreation areas.

Advocates of national recreation areas argued that these new inclusions did not represent a dilution of park ideals. They reasoned that the original legislation for national parks had said that public use was a crucial element in national park purpose. Since NRAs were located near urban areas, they promoted public use through their convenient (and less expensive) access.

Recreation areas were therefore a solution to concerns about access and recreation for the American public as well as protection against commercial, industrial, or residential development.

Advocates further argued that for nearly a century parks had functioned as federal subsidies for middle-class Americans who could afford the vacation time to visit and use the parks.[50] They pointed out that there was no federally supported recreational opportunity available to the urban poor, and they emphasized that the original legislation mandated parks for all the people, not just the affluent. Thus recreation areas helped the national park system expand its role in a democracy by increasing access. But soon after the establishment of NRAs, these sites were widely regarded as areas of less merit than nature sites. The word *recreation* has become a pejorative term in the lexicon of many Park Service professionals.[51]

In response to a century of increasing complexity and diversity in the park system, and perhaps in response to concerns about the value and appropriateness of the newest additions, in 1971 Congress passed the General Authorities Act, an amendment to the National Park Service Organic Act of 1916. This legislation specified that all units administered by the National Park Service were part of the same system. It stated emphatically:

> Congress declares that the national park system, which began with establishment of Yellowstone National Park in 1872, has since grown to include superlative natural, historic and recreation areas in every major region of the United States, its territories and island possessions; that these areas, though distinct in character, are united through their interrelated purposes and resources into one national park system as cumulative expressions of a single National heritage; that, individually and collectively, these areas derive increased national dignity and recognition of their superb environmental quality through their inclusion jointly with each other in one national park system preserved and managed for the benefit and inspiration of all the people of the United States. (PL 84-825, preamble)

The implied message of this statement was that all units within the park system form a whole, with each unit occupying a niche in a coordinated system. However, people inside and outside the Park Service continued to rely on categories (such as national battlefield or national monument) to describe the features of a particular site.[52] This legislation is revealing. It highlights the fact that there was some confusion over national park ideals. Supporters of this legislation felt it was imperative to confirm the wholeness or integrity of the system itself while simultaneously recognizing the value of each individual unit and appreciating the diverse ways in which each park contributes to the system. This could have been addressed with a Park Service memorandum, but it is telling that this issue was deemed important enough to warrant federal legislation. The General Authorities Act also illustrates the fact that the parks and park ideals have been in continual evolution, at times punctuated by needs to clarify purpose and reinterpret park ideals.

Despite the controversial nature of the NRAs, there was significant political support for them, and this political support led to the establishment of additional parks and recreation areas. Following his success in securing the GGNRA legislation, Philip Burton sponsored a larger legislative package proposing new local, regional, and urban national parks. In 1977, he pushed through Congress the largest single legislative package in national park history.[53] Burton's legislation, the National Parks and Recreation Act, added fifteen new units to the national park system. It also marked the apex of its expansion.

Political Battles over Specific Parks or Park Policy

The four park movements have been a powerful force in an evolutionary process within the national park system. The meaning of national parks has also been defined by a series of political battles over specific parks or policy. Often these struggles set precedents or defined new ideals. In some instances political struggles were incidental but nevertheless influential on the national parks. The history of the parks is filled with numerous battles; I intend to introduce only a few examples in order to demonstrate that a political contest over a specific park often has an impact on the entire system.

The creation of Everglades National Park in 1934 was not part and parcel of a park movement; nevertheless, the creation of this park greatly influenced national park principles. The Everglades was devoid of monumentalism; its topography hardly suggested drama or splendor. The battle for this new park propelled conservationists to articulate the importance of the new science called ecology. The Park Service and park advocates argued that the Everglades was worthy of inclusion in the system because of its unique and fragile ecology. The successful battle to secure congressional authorization thus confirmed a new commitment to protect and preserve an ecological system. Everglades National Park became the first national park based on a biological perspective and conservationist ideology.[54] Support for the new park indicated that romanticism, while not entirely dead, had made way for an emerging scientific rationale of conservation ecology. Thus the creation of this new park, at the time considered an anomaly, symbolized the emergence of new ideals for national parks. The lasting implications of the addition of this park were profound. It marked a new direction in park management policy, and since its inclusion the Park Service has elevated conservation ecology ideas as a guiding principle in park management policy.[55]

Other political struggles also set precedents for park ideals. One of the more influential struggles, and one that environmentalists still lament today, was the battle over Hetch Hetchy in California's Yosemite National Park. In 1901, San Francisco city developers petitioned the Department of the Interior to use Hetch Hetchy Valley for a dam and reservoir, thereby providing a permanent supply of fresh water for a growing city of more than 500,000 people. This proposed development project threatened to set a new precedent for the entire park system. Prior to the Hetch Hetchy battle, schemes to exclude lands from national parks had been limited to the outside edges of the reserves. Hetch Hetchy Valley, however, was not on the fringe of Yosemite National Park but centrally located. The political struggle over Hetch Hetchy was a struggle over the integrity of the Park Service: if the inner sanctum of Yosemite could not be protected in perpetuity, no national park could be considered safe.[56]

Park preservationists, led most vocally by John Muir, argued that the valley rivaled the splendor of Yosemite and thus should remain protected from development. But their fervent arguments did not sway Congress. In 1913, Congress voted to permit San Francisco to begin construction of the dam and to flood the valley floor. Today, Hetch Hetchy Valley lies under water. The loss of Hetch Hetchy had a profound impact on conservation and environmental organizations. Many organizations established permanent lobbying groups to fight similar threats; these groups proved to be important in future struggles.

The battle over Hetch Hetchy was not the first struggle to protect parks against divestiture or development, nor was it the last. Since their inception, many parks have been under threat—from hunting, logging, or mining interests and, more recently, from the strain of massive tourism and pollution. Development projects have periodically threatened the integrity of parks and generated heated public debate. Such projects have often resulted in numerous legislative controversies. One of the most notable involved the proposal to deactivate or decommission Jackson Hole National Monument in Wyoming. In 1943, Congress was persuaded that opportunities for economic development in timber, mining, and grazing outweighed the value of the

Jackson Hole National Monument to the American public and voted to abolish the park. Eventually this legislation was vetoed by President Franklin Roosevelt. This struggle highlights two important points. First, political elites have not always acted to protect and preserve the parks. Second, attempts to decommission national park property have occurred. Although the permanence of the national park system is taken for granted, more than sixty areas have been decommissioned; some were transferred to state or local government, and others were returned to private ownership.[57] Representative John Duncan's attempts to dissolve the Presidio were not without precedent.

In addition to overt political contests over specific national parks, the results of broader national political debates have impacted the parks. For example, in some instances federal legislation not intended to change or modify the park system itself has affected Park Service philosophy and policy. The Wilderness Act of 1964 and the National Environmental Policy Act of 1970 set new requirements for environmental protection and mandated that all federal agencies (including the Park Service) implement the new policies. The flurry of environmental legislation during the 1970s—the Clean Air Act, the Clean Water Act, the Endangered Species Act—also redefined the role of the Park Service in protection and preservation. Many of these new laws asserted that the guiding principle for the national parks should be preservation and interpretation of natural landscapes and ecosystems; this directly affected Park Service management policy.[58]

Public debates about the condition of parks and direction of park policy have also generated political struggles. Often these debates are results of Park Service reports or agendas for the future. In 1980, the National Park Service responded to a congressional request for a report on the state of the parks.[59] The report was to identify and describe threats that endangered the parks and to discuss the broad spectrum of problems facing the parks. The Park Service was charged to discover which threats and problems would damage park resources or seriously degrade important park values.[60] The report was disheartening. The parks, it concluded, suffered from a wide variety of threats: internal threats such as overcrowding and overbuilding; external threats such as urban encroachment and air and water pollution. The Park Service concluded that the 1990s would require significant action and an infusion of federal funds in order to protect and preserve all the parks, but especially the nation's crown jewels.

Between 1960 and 1990, the Park Service tripled its acreage, and the number of visitors to the parks increased from 80 million to 257 million.[61] Many felt the Park Service stood at a crossroads.[62] In 1991, the National Park

Service celebrated its seventy-fifth anniversary by holding a symposium at Vail, Colorado. The event brought together hundreds of experts, Park Service employees, and other interested parties. The symposium was designed to address issues of critical importance to the Park Service, to review challenges and strategies for the twenty-first century, and to reconcile the broad range of challenges to the Park Service mission.[63] It proved to be an unprecedented display of bureaucratic introspection.[64] The result of the symposium was a 137-page document referred to as the *Vail Agenda*. While acknowledging that the Park Service's purposes had evolved significantly in the first seventy-five years, the document reaffirmed that the national parks were places that symbolize a defining time in American history.

The symposium outlined new objectives and introduced many suggestions. Of note was the suggestion that the Park Service organize an office dedicated to policy. This suggestion was in part a reflection of the fact that the Park Service has no central office to handle national politics, and as a result each park battle is fought as a separate issue rather than as part of a broader park policy. James Ridenour, National Park Service director from 1989 to 1993, commented that by the mid-to-late 1980s "The Park Service had become a captive of micromanagement by Congress. . . . We weren't anticipating; we were reacting. . . . the creation of a new strategic planning office was a step in the right direction."[65]

The suggestion to create a policy office was also a recognition of an inescapable conclusion. If we accept that the ideals behind the national parks as well as the national parks themselves are cultural expressions, then both the ideals and the parks are social constructs. Since the National Park Service falls under congressional jurisdiction, the creation and redefinition of national parks take place in an *inherently* political context—the political process unfolds on the national political stage. National parks are, and will continue to be, subject to the political climate as long as Congress retains its jurisdiction. Congressional debate over the future of the Presidio was not a political aberration; it did not magically surface in 1993; it emerged from a political and cultural context.

THE DYNAMIC EVOLUTION OF NATIONAL PARKS

The history of America's national parks is an outcome of America's long and complex intellectual history, one full of debate, diversity, and contradictions. The national parks and the national park system constitute a dynamic mosaic crafted anew by each successive generation or park movement yet reflecting the array of political, social, economic, and technological forces that have shaped American history. For all its history, though, the national park system has been an improvisation.[66] A series of political conflicts and park movements has altered in some way the management, the acquisition, and the meaning of parks in America. Each has introduced or shifted an ideological emphasis, introduced new standards and practices, shown the vulnerability of parks to a broader political process, or set alarming precedents. The absence of an agreed-upon standard by which to measure a park's value or contribution has left an indelible legacy within the Park Service. In the early years, many Park Service professionals were uncomfortable managing historic sites (many still are); in later years, many were equally reluctant to take on the responsibility of urban NRAs.[67]

As our understanding of nature has changed, so too has the role of national parks. Parks have served as a barometer of society's changing attitudes and perceptions, and ideas about what parks are, or should be, have evolved through the years.[68] Dominant or traditional views have been challenged; ideals have become less romantic and more ecologically oriented; preservation now exists, albeit in conflict with a heightened commitment to recreation; and finally criteria for new parks have been broadened, allowing an appreciation for the many ways in which people can experience and appreciate the out-of-doors. As a result, our ideas about national parks, and about the Park Service itself, have been subject to periodic adjustments; in turn, these adjustments provoke deep philosophical divides over the meanings of parks. The Presidio is one such illustration.

Thus there have always been conflicts about the purpose and ideals of national parks, there have always been worries about inadequate funding, and there have always been threats to the integrity of the system itself. The history of the Park Service has been characterized by the growing complexity of its mission, changing agendas, new policies, and the shifting of park ideals. I believe this evolution can be seen as a dilution of the original concept of national parks or as a broader effort to articulate the many possible relationships of humans to their environment.

There are two ways to view the present-day park system. First, some see the system as a collection of sites that unevenly represent America's natural and cultural heritage. This view leads to the conclusion that the system exists as a hierarchy in which some parks are "more equal" than others and that the national park ideal has been compromised.[69] The second viewpoint suggests that the national park system is more than merely the sum of its parts. This view leads to the conclusion that no park is better than or inferior to others, only different, and that simply by being included in the system, each park unit has achieved a measure of national significance. The latter view also recognizes that each park is an expression of its time, a statement by one generation to those who come later; thus efforts to devalue or divest the system of a given park are just a reflection of changes in social values, economic cycles, or perspectives.[70]

The overriding purpose of the National Park Service is not only to protect resources of significance to the nation but also to convey the meanings of those resources to the public.[71] Our understanding of national parks is as preservers of "the historical, cultural and natural foundations of the Nation," and so it is logical that they constitute "natural and cultural resources that contribute to the nation's values, character and experience."[72] Parks are not just places of recreation and entertainment; they are not just scenic wonders; they are not merely memorable; they are *places of meaning*. Ideally, parks constitute (both individually and collectively) the best of our natural and cultural resources. These are places of remarkable value that are "too special, too precious ever to be reduced to private ownership and exploitation, but should be retained for the enjoyment and inspiration of all people."[73] In contemporary American society, national parks are more than just beautiful places, more than history and heritage, touching instead something deep in the human spirit.[74]

After a century of change and expansion, one thing seems clear: there can be no single overriding ideal, no single criterion, that defines natural or historic significance in a system defined by incredible diversity and complexity. In fact, any attempt to impose a single purpose or a single philosophy would be inadequate and inappropriate.

Nature, Culture, the Park Service, and the Presidio

The history of national parks and the national park ideal is important to the story of the Presidio for several reasons. First, it shows that our understanding about national parks is dynamic. The inclusion of any new type of park occasioned a large and often divisive debate about appropriateness, about criteria, about worthiness. Second, the legacy of the crown jewel syndrome has meant that people (especially those in Congress) do pay attention to the position in the hierarchy when considering federal appropriations. The status of "recreation area" implies a marginal importance to the system. Third, as a consequence of the deeply embedded preference for scenic monumentalism/crown jewel worship, the establishment of national seashores, lakeshores, and urban recreation areas crystallized public debate over the accusation that these parks were less worthy and would dilute park values. Urban NRAs in particular have been burdened with the perception that they are more limited in their scenic impressiveness and hence less deserving of designation. They are not crown jewels. The legacy of the nature/culture hierarchy, the marginalization of the recreation areas, and a lingering antiurban park bias (which in particular affects the urban recreation areas) have been subtle but powerful forces. Together, they have had an enduring impact on the status of recreation areas, how the Park Service administration views these areas, and how Americans view and value the recreation areas.

Nowhere is this fact more apparent than in the words of former Park Service director James Ridenour (who was in office for the first several years of the Presidio planning). In 1990, he boldly stated that there were many "undeserving" parks in the national park system, and he disparaged their presence as "thinning the blood" of the Park Service:

> Our greatest problem was the creation of too many parks that were not truly of park stature.... Members of Congress have blatantly disregarded standards that have been traditionally used in evaluating the creation of new national park units. They have turned "pork barrel" into "park barrel," and they are thinning the blood of the NPS. Many of the units being voted in by Congress are not worthy of national recognition but get voted in anyway. That thins the quality of the system and puts additional financial demands on an already badly underfunded program. We are not taking care of the Grand Canyons, the Yellowstones, the Everglades ... *while we spend hundreds of millions of dollars on what can best be described as local or regional economic development sites.* To me, that is thinning the blood of the system [my emphasis].[75]

In subsequent articles, Ridenour asked readers to reflect on recently added parks and to compare them to the crown jewels. The term *park barrel* became a popular way of saying that the once high standards for national parks had been compromised in the interests of local economic development projects (such as recreation areas and other urban units).

Similarly, in 1985, Ronald Foresta in his book *America's National Parks and Their Keepers* declared that "the entire urban wing of the National Park System, was, in essence, an experiment, and the [NPS ought] to ... write off failures when the time comes to do so."[76] Not everyone in the Park Service felt this was an accurate assessment. Dr. Robin Winks, former chair of the National Park System Advisory Board, visited 333 units of the park system. He observed that only a few of them did not make the grade and concluded that "this is hardly a decline in quality."[77] But Ridenour's critique, though not new, was authoritative.

Who was the architect of this park barrel trend that Ridenour criticizes? None other than Philip Burton, who had sponsored the creation of the GGNRA in 1972 and later the National Parks and Recreation Act of 1977, the legislation that added fifteen new units to the national park system.

In part the "thinning the blood" accusation is a response to a perceived trend to hold the National Park Service budget steady, despite new additions and responsibilities. It is true: while Congress has been quick to endorse new parks and projects, it has been hesitant to give the parks the money for upkeep of the facilities. There is an estimated $2 billion backlog of maintenance and repair projects. Park advocates believe this is because routine repair projects are not where people will notice—plumbing, wiring, and painting do not lend themselves to dedication ceremonies and ribbon cutting.[78] Congress considers the creation of new parks a great achievement (a park in every district is good for reelection); but approving routine but crucial infrastructure maintenance has little glory. It is an interesting paradox. In recent years, Congress has been among the most vociferous critics of the Park Service for its backlog of maintenance and repairs, an ironic charge because Congress has withheld approval for maintenance appropriations or has directed general appropriations for projects other than maintenance. Ridenour was correct in perceiving that an increasingly hostile or reluctant Congress might mean continued budget cuts. This concern has proved to be an accurate prediction, for as we saw in the Presidio battle, Congress has become increasingly fiscally conservative. It would seem, then, that concern over Park Service finances lay at the heart of Ridenour's accusation. He continued:

> Good and not-so-good park sites came into the system under Burton's time in the chair, but none that have impacted or will impact the operations of the NPS in a more substantial way than the establishment of the Presidio of San Francisco, a former military base.
>
> The Presidio is truly a great property, but the problem is trying to find the right use of it that will allow for a park-like setting without costing the taxpayers huge sums of money for its operation. I was told that the NPS might expect to spend $40 million a year just for operating expenses. Compare that with the annual operations budget at Yellowstone, which is about $18 million. I hope the planners and politicians can make it work. I especially hope they can make it work without bankrupting the rest of the sites in the NPS.[79]

This is hardly a ringing endorsement from a former Park Service director. It is possible that had Congress funded the Park Service adequately, the issue of "thinning the blood" would never have emerged and Ridenour would not have felt compelled to compare the operating budgets of the Presidio and Yellowstone. Many Presidio supporters, however, felt that the accusation had less to do with financial pressure and more to do with a philosophical divide over national park ideals. One observer explained:

> At the same time more innovative units of the Park System were introduced, dollars got tighter. Some Park Service administrators made the assumption that the new parks were the cause of the financial strain. The intellectual way to handle that, without appearing too greedy, was to say that these new parks didn't belong in the system, rather than being honest and saying, "I'd really rather have that money for Yellowstone."[80]

Ridenour, like several of the congressional opponents to the Presidio, implied that the Presidio is not of the same merit as Yellowstone and that it is, like many other urban parks, really a local economic development project in disguise. Coming from a former Park Service director, this statement certainly attests to the enduring debate about the worthiness of urban parks in the Park Service. Rettie writes that this ideological position "reflects a dangerous, unfortunate and legally questionable disregard for the integrity of the national park system."[81] The 1991 *Vail Agenda* affirmed that while all units contribute to public values, the ways in which these contributions are made varied. Each park thus contributes its unique identity to the national character. In other words, the strength of the national park system is in its diversity. So it could be argued that the Statue of Liberty, the White House, and the Presidio are just as crucial to our image of ourselves and our identity as Yellowstone and Yosemite are.

Many people involved with the Presidio at the local level—GGNPA staff, Presidio Council members, community organizers, and many Park Service GGNRA staff—were dismayed that notable Park Service figures such as Ridenour believed the Presidio project would thin the blood of the system. "It was very difficult," said one observer, "to have high-profile people in the Park Service talk about 'real' national parks and insinuate that there were 'other' parks like the Presidio."[82] Some were disheartened that not only Congress but also some in the Park Service questioned the worthiness of the Presidio. Brian O'Neill, GGNRA superintendent, felt there was significant resistance about the Presidio from within the Park Service but noted that "those in the Park Service who still question the Presidio's worth haven't come out here [to San Francisco]."[83]

As a counterpoint to Ridenour, however, William Reilly, in his testimony on behalf of GGNPA before a Senate subcommittee, summed up the status of recreation areas and alluded to the future direction of the national parks:

> The great urban parks—Santa Monica, the gateways in San Francisco and New York, Chattahoochee, Cuyahoga—all of them were the subject of a debate, and the question that was asked about them was are they worthy for inclusion in a system that also has to support the crown jewels, the Yosemites, the Grand Canyons, the Yellowstones and the Everglades?
>
> I think that on all, really, of the relevant grounds, whether natural resource, historic and cultural significance, and value, archeological characteristics, the Presidio qualifies as a crown jewel. I think it is the most significant land-use opportunity in the United States. It already receives more visitors than Yellowstone, Yosemite and the Grand Canyon together.
>
> It is, however, a very different kind of park. I do not think we are likely to see in the future park opportunities of the sort that we confronted in Alaska, for example, where we have substantial areas of still pristine undeveloped land. The new parks will very likely have characteristics of development, of some maturity possibly, as in the case of the Presidio. . . . There is a lot of history there. This is a mature facility. But it is an absolutely outstanding facility.[84]

In the case of the Presidio, the ideological debate between nature/culture, the crown jewels/the rest, moved no closer toward resolution. Indeed, it remains a thorny issue.

The Golden Gate National Parks: What's in a Name?

Consider again GGNPA's campaign to rename the Golden Gate National Recreation Area, in large measure to "clearly acknowledge this park's status as a unit of the national park system, clarifying this issue for many who are confused by the term 'recreation area.'" Nomenclature is no small issue, especially for NRAs that have been referred to not as national parks but as recreation areas or even as playgrounds. Such is the power of a label to diminish the value of particular parks. As Greg Moore, GGNPA executive director, commented, "It's too bad the Park Service developed all these designations like recreation area, national historic site, or national seashore rather than calling each unit, regardless of its natural or cultural emphasis, a national park. The national parks are not just crown jewels; they are places that contain interwoven parts of our history."[86] Indeed, the GGNPA name campaign attests to the limitations and negative consequences of such categories.

GGNPA felt the need to remind the public (and perhaps the Park Service) that the Golden Gate National Recreation Area was a *national park*. The name change was clearly an attempt to capture the aura, mystique, and prestige associated with "real" national parks and to move beyond the bias against urban parks. It was also an attempt to make individuals in an increasingly diverse population stakeholders—emotional, intellectual, and economic—in the national park system. The campaign to confirm the status of the GGNRA within the national park ideal also mirrored the Park Service's efforts to define the system for the next century. Ridenour was correct when he asserted that the Presidio would impact greatly on the national park system. But it is too early to tell what this impact will be.

CHAPTER 9

Transforming the Presidio: An Analysis and Some Conclusions

This book has constructed a biography of one place, the Presidio. I have examined the vision for the Presidio's transformation, discussed how an enthusiastic and committed citizenry was crucial to both the planning process and the political struggle in Congress, and speculated about the reasons why this vision faced (and continues to face) opposition. This project has shown the way in which nature in the form of national parks is a social and political construct.

It is difficult to make definitive statements or sweeping conclusions, because the Presidio is only in its first years as a national park and another twelve to fifteen years must pass before the Grand Vision is fully realized. It is also premature to reach conclusions about the impact the Presidio project will have on other national parks or national park policy, because the proposed Grand Vision and the Presidio Trust public-private partnership constitute a solution to the problems associated with a highly complex and specific place. However, several interesting aspects of the Presidio and the process of creating a national park need to be integrated in a conclusion. In particular, I want to look at five themes: the making of the Presidio; casting a critical eye on the Grand Vision; the politics of parks; the relationship of city and park and "constituency building"; and the exploration of the nature/culture and city/park dualisms.

The Making of the Presidio and Special Places

A historical geography of the Presidio reveals deep, rich layers of human activity, as well as transformations of the Presidio's meaning and physical environment. The Presidio began in 1776 as a Spanish garrison on the fringe of New Spain's western frontier. For the next 218 years, it remained an operating military post under Spanish, then Mexican, then American flags. Throughout its history, however, the Presidio has been inextricably tied to the city of San Francisco. In turn, San Franciscans have perceived the post to be part of the city.

The uniquely open character of the Presidio post has meant that many San Franciscans have used and cherished the Presidio as an urban space. People have mobilized to preserve and protect it from urban encroachment or development schemes. One of the most significant efforts occurred in the early 1970s, when citizens organized the People for the Golden Gate National Recreation Area and convinced California congressman Philip Burton of San Francisco to sponsor legislation creating the Golden Gate National Recreation Area. Burton shrewdly included the Presidio in the GGNRA.

The creation of the GGNRA in 1972 can be considered a type of social movement aimed at "greening the city," or bringing together nature and culture in the city.[1] The movement to create and preserve urban parks and open spaces in San Francisco did not end with the GGNRA's establishment. Numerous individuals with the support of an array of organizations established several nonprofit organizations, including the Golden Gate National Park Association, People for the Parks, and People for the Presidio. Several serve as members of the GGNRA Advisory Commission. These organizations oversee the administration of the GGNRA and act as watchdogs over these protected urban open spaces.

CASTING A CRITICAL EYE ON THE GRAND VISION FOR THE PRESIDIO

> **There are occasions in which an opportunity is so palpable, a time for action so precisely apparent, and a place so right that even our contentious species cannot and will not lose the chance to achieve a grand result.**
> —Roger Kennedy, Preamble to the Presidio Plan (1993)

The 1988 decision to close the Presidio thrust the post into a new role: that of a national park. The creation of the Grand Vision was a result of a three-year planning process that involved significant citizen input and Park Service guidance.

In the history of planning in the United States, so-called grand visions have often marked important transitions. Such comprehensive master plans have defined new planning ideologies. Daniel Burnham's Chicago, the

American versions of Ebenezer Howard's Garden Cities, the bioregional communities of Patrick Geddes, and postwar planned communities such as Levittown and Irvine relied on comprehensive planned visions. These grand visions bequeathed to the planning profession and students of the city an important intellectual legacy because they focused on space not as it was but *as it could be*. Similarly, the Presidio's Grand Vision imagines an urban national park as it might best benefit the larger community; it strives to transform an old military post into an urban national park by imagining a new role for both the Park Service and the Presidio. This seems worth emphasizing. Civic leaders and business elites might have worked together to divest the Park Service of Presidio lands in order to promote residential or commercial development. They could have argued that the community would benefit more from the developers' fees, tax revenues, and increased property values in surrounding neighborhoods that would have resulted from such development. They didn't. Rather, they worked together to support the Presidio plan and the trust legislation. Therefore community efforts to protect the Presidio can be seen as an act of social justice because this increased public access as a national park to a much greater degree than a residential development would have, while costing more money (federal tax dollars) than it might have generated as a commercial development.

The urban geographer Peter Hall has situated the quality of planning along a continuum ranging from a "planning disaster" to a "place of tomorrow."[2] Similarly, the ideology of planners themselves can be situated along a continuum. At one end are pragmatists, who look at space as it is, accept the status quo, and attempt to create efficient, scientific, but often conceptually narrow plans. At the other end are utopians, who strive to rearrange geographic and social relationships and imagine space as it could be. The Presidio plan and the ideology of the Presidio planning team fall somewhere in between, although the plan itself asserts utopian ideals of a "park for the 21st century."[3]

The Grand Vision is a plan broad enough to allow the Presidio to be many things to many people. Yet it also establishes a common community purpose in the exploration and education of environmental and social problems. Advocates hope the Presidio will contribute to the city, the Bay Area, and the nation as a place of environmental leadership and public service outreach. This plan envisions a community in greater contact with nature, exploring ideas about the environment and the meaning of national parks. This is a profound concept. Many social critics attribute our social and environmental problems to the increasing separation of humanity and nature.

The Presidio's Grand Vision, however idealistic, represents an attempt to reconcile people and the environment, nature and culture, cities and (national) parks. Although we should be wary, even critical, of the hyperbole associated with the term *grand vision,* the plan strives to transform a space into a public place that matters to and benefits the community.

IMAGINING THE FUTURE

Good planning has always been involved with images of the future. Effective planners provide workable plans and supply a vision of a better tomorrow.[4] Images of the future are of critical importance in influencing which of the possible futures becomes the present reality.[5] Planning can expand civic consciousness, raise new standards of design, propose new ways to consider space. "As long as a society's image of the future is positive and flourishing, the flower of culture is in full bloom. . . . without new images of an ideal future to guide man's striving, civilization is doomed."[6] A good plan anticipates the future, not as a glorified extension of the present but as the formulation of trends and ideas only beginning to emerge.[7] Of all the proposed alternatives, of all of the possibilities for the future of this space, the Park Service's Grand Vision transcended the requirements of the Burton legislation. In so doing, the Park Service *expanded* on its traditional mission by calling for the Presidio's buildings to be transformed into centers of research, learning, and education about the environment and about social and cultural conditions.

The Presidio plan is more than a product; it is also a process of transformation and change—for both the city and the Park Service. Individuals and groups who participated in the planning were changed by their participation because they were encouraged to envision the Presidio *as it could be.* They helped to shape and give meaning to the role of the Presidio as a national park. The Park Service, too, was changed by both the planning and the political process. The 1991 *Vail Agenda* established new objectives for the Park Service over the next several decades. One objective challenged it to expand and deepen its role as the guardian and steward of, and educator about, natural and historic resources. The proposal to restore landscapes, renovate buildings, and establish a global environmental and education center expands the role of the National Park Service and perhaps lays the framework for future urban parks. Under the Grand Vision for the Presidio, the Park Service has stretched its imagination, reemphasized its role in education on environmental stewardship. The Grand Vision also reflects how persuasive and influential concern for the environment and environmental quality has become in our society. This concern affects the way people think of place and the uses of space, including parks.

Ironically, even after the realization of the Grand Vision, the Presidio's form is not expected to change dramatically. When asked to imagine the Presidio in the year 2010, many observers believed that the space itself will look similar to the way it looks today.[8] Differences will be both subtle (fewer parking lots, fewer nonhistoric buildings) and abstractly profound (more community activities and educational opportunities, more buildings occupied by a diverse collection of organizations, a greater sense of community and purpose). A greater sense of purpose will emerge in large part from the global environmental center. *The "transformation" of the Presidio has less to do with radical changes in infrastructure, the physical landscape, or buildings, and more to do with mission, purpose, and meaning.* The Presidio's plan is thus a cultural expression of the importance of addressing environmental issues as social issues.

Although it may be premature to judge the Presidio plan or the Presidio as a park/place of tomorrow, both the planning process and the Grand Vision make a strong case for imaginative, inclusive planning. Nevertheless, the vision, however noble and inspiring, must contend with political forces.

Successful plans need public participation and endorsement. In this case it was not only wise to have public involvement but was absolutely critical, given the park's urban location. Public involvement in the Presidio's transformation was the principal mechanism by which Park Service planners communicated their proposals, policies, and intentions to the city. Many other Park Service plans have languished for years without public endorsement, their planning process full of tension, vociferous disagreements, name calling, table pounding, and fist waving.[9] It is altogether possible that the Presidio plan, because of its high visibility within its urban context, could have met with widespread resistance or lawsuits filed by disgruntled special interest groups. One reason this did not occur was that the Park Service had encouraged the public to dream big, to stretch their imaginations, and to participate. This resulted in widespread endorsement of the Presidio plan. Although this strategy was effective in building the local constituency for the Presidio park, it proved to have negative consequences elsewhere.

THE POLITICS OF PARKS

There is perhaps no better illustration of the way in which social/political processes define a park than in the struggle over the Presidio Trust legislation in Congress.

Some national park advocates admit that congressional meddling and petty politics have led to inadequate, even inappropriate units.[10] It is also my contention that congressional meddling and petty politics threaten to undermine the future of national parks, hindering the opportunity of parks to provide environmental and educational leadership—in short, to limit the possibilities of *what* and *where* a national park can be.

There were many reasons behind congressional resistance and hostility toward the Presidio plan. Foremost was the issue of cost. The Presidio plan was grand in its vision, and it was also grand in its cost (it is important not to underestimate this factor). One of the major complaints about many Park Service plans is that they are unrealistic—too grandiose or too expensive or too ambitious. Planners create the visions; then the accountants tally the needed expenditures. For example, the estimated cost of the plan was not discussed until page 112 of the Draft Management Plan, and then merited only a vague two-page discussion and a brief summary table placed in the Appendix. The Presidio's Grand Vision was certainly expensive. At a cost of $24 million per year from the government over ten years, the Grand Vision relied on a considerable amount of federal money, which ultimately made it vulnerable to political criticism. Some in Congress never saw past the overall $660 million figure and criticized the plan for being economically unrealistic. "It's a great vision with potential, but the price tag was just too much," explained one frustrated advocate.[11]

One of the problems caused by the vision-first, costs-later approach in the Park Service was to set unrealistic expectations about the kind of community that would reside at the Presidio.[12] Many San Franciscans envisioned the Presidio park as a nonprofit campus and were not prepared for the fact that the enormous annual operating expenses would entail some form of private business participation. When Congress inserted financial restrictions into the legislation, and the full extent of the need for private business to help support the Presidio became understood, there was some community backlash. Generally, this took the form of the *Bay Guardian* editorials, although others in the city became concerned about "too much business in a national park."[13] One *Bay Guardian* editorial decried the "privatization" of the Presidio and labeled the Presidio Trust "Presidio Inc."[14] Unfortunately for Presidio advocates, several members of Congress seized on this criticism as proof that the city of San Francisco was divided over the Presidio plan.

More importantly, however, the expense of the Presidio's Grand Vision quickly became enmeshed in a complex national political battle. In part this was due to the visions-first, costs-later culture inherent in Park Service planning, which meant there was little incentive for planning and budgeting to appear "in sync."[15]

The cost of the Presidio plan was first publicized in the early opposition led by Representative Duncan, who attempted to curtail Park Service funding for the Presidio. As a result of these early challenges, nearly everyone in Congress, the media, and the American public became aware of the enormous costs of the Presidio. (In 1994, news anchor Peter Jennings highlighted the Presidio plan as an example of wasteful government spending in NBC's *Fleecing of America* series.) In previous decades the Presidio's cost might not have sparked such a debate. But the agenda of the 103rd and 104th Congresses was evolving in a new direction. There was growing concern and discussion about the national debt and the federal budget deficit. By 1993, a clear trend in Congress had emerged: fiscal conservatism. It encompassed both political parties. Amid this new political climate were public debates about welfare reform, health care reform, Social Security reform, and overall budget reductions. All federal agencies were expected to face lower increases in their annual appropriations; some would even incur reduced budgets. The National Park Service was no exception.

As a consequence of the trend in fiscal conservatism, Congress issued the Park Service a challenge: how could it pay for this $24-million-a-year park with no significant additional overall budget increase and a very sizable backlog of repair projects in other parks? From where would the Park Service take the funding and the staff? In other words, how could the Park Service make the Presidio work without compromising other parks? Many, including even Presidio advocates such as William K. Reilly, believe the Park Service was never able to meet this challenge, in part because Congress chose not to believe the agency's response.[16] It is worth noting that this challenge was bipartisan. It arose initially from the Democratic-controlled 103rd Congress and remained an issue in the Republican-controlled 104th Congress. It was the Presidio's most obvious stumbling block.

The political battle—the numerous debates about, filibusters on, and compromises to the Presidio Trust legislation—illustrates just how politicized the national parks had become. In the case of the Presidio, the park's future became a highly contentious and politicized issue to which all sorts of other agendas (and bills) were attached. The Presidio Trust, in its penultimate draft form, represented debate, negotiation, and compromise. It was the concrete manifestation of the social/political creation of parks. Many worried that it signaled a heightened level of politics for parks. This remains to be seen, but my considered opinion is that it did.

The political struggle for the Presidio required the advocacy of the residents and organizations of the city. In fact, the city itself played a vital, if not the primary, role in securing the Presidio Trust legislation. GGNPA had to learn how to operate in and understand the federal legislative process, and it faced a significant public relations challenge. Greg Moore, GGNPA executive director, explained how the political battle changed the organization:

> We never used to worry. We were just here to help. Now we know to be proactive. *I think the GGNRA has come to depend upon us, rather than the Park Service, to help them in that way.* And that's very critical. By the time the needs of the GGNRA get to the Washington office, they are competing with the needs of 360 other parks. So on the positive side, it brought visibility; it brought wonderful volunteers; it brought a new skill base. I can't imagine now our organization not having a government relations function [my emphasis].[17]

I highlight this comment because it demonstrates how very important nonprofit organizations were in the Presidio struggle.

The implications of the Presidio struggle on the wider national park system remain to be seen. It does appear that the current rhetoric of fiscal conservatism will put even more financial pressure on the park system. Innovative proposals for other new parks or creative management strategies for older parks are likely to face resistance or outright hostility.

It is a logical outcome of this new fiscal conservatism that all federal bureaucracies face budget challenges. However, the hostility toward the Presidio plan also raises the larger issue of *spending priorities,* both in the National Park Service budget allocations and the larger federal budget priorities. As historian Alfred Runte has succinctly and eloquently remarked, "history may think it strange that Americans could afford $4 billion fighter planes while a park was considered beyond its means."[18]

"IT'S NOT EASY BEING GREEN"

It is unlikely that Kermit the Frog had the state of national parks in mind when he (through Jim Henson) immortalized this song, but it seems an apt, succinct appraisal of the political opposition the Presidio plan encountered. No doubt the primary justification for congressional resistance lay in the cost of the plan. However, it is also my contention that some congressional resistance can be attributed to a more subtle criticism: a growing mistrust of the Park Service and the political agenda of some in Congress to assert more discipline and control over the parks. In recent years the Park Service has come under fire. It has been accused of inefficient management. It has been criticized for a large maintenance backlog. These may be fairly accurate assessments, but the Park Service is no different from any other bureaucracy—all are inefficient to some degree. Although congressional criticism appears to have intensified, many Americans believe the national parks are among the best ideas Americans ever had, and the Park Service is one of the most respected bureaucracies in the nation.

Although there have been other federal public-private partnerships, the concept underlying the Presidio Trust had never been attempted in a national park, so there existed no blueprint for its implementation. A great deal of education had to occur; numerous questions had to be answered. Every aspect of the trust was fought over, and because it is a federal corporation, the charter is the legislation. "We also learned that you always have to deal with the minority because they become the majority. You can't assume that the same people will stay in power. You have to educate and work with everyone in Congress. On the advocacy side, you have to build the widest possible coalition. That's the only way to get a project like this through," said Craig Middleton, director of government relations for GGNPA.[19]

The Presidio Trust was a response to previous criticism that the Park Service needed to find constructive mechanisms for developing new and innovative approaches to park management. Many thought the Presidio Trust partnership was just that. It is, therefore, ironic that congressional criticism of the Presidio Trust legislation was led most vocally by Republicans, many of whom had strongly promoted an agenda of deregulation. The rhetoric of deregulation promotes less government and more local control; it encourages public-private partnerships. In the Presidio battle, just the opposite occurred: more government control—in this case the Park Service was "out" and the Congress was "in."

It is also my contention that congressional resistance to the Presidio can be traced back to the lingering perception among many that urban recreation areas are not legitimate national parks and are less deserving of large operating budgets than the traditional parks. This attitude was expressed by members of Congress and by certain Park Service officials who questioned the Presidio's national significance or its worthiness. More than once, the Presidio budget was compared to that of Yellowstone and Yosemite. It is, of course, difficult to confirm with anyone in the Park Service or Congress that this criticism or challenge underlay any of the debates about the Presidio. But local Presidio advocates—in GGNPA, on the Presidio Council, and among

Park Service GGNRA employees—perceived this to be a real issue. Perhaps the way to understand this legacy of the crown jewel syndrome or scenic monumentalism is to situate it as a question of geography.

BUILDING A CONSTITUENCY FOR THE NATIONAL PARKS: THE CITY AND THE PRESIDIO

In 1993 and 1994, the rhetoric in congressional debate focused on the question "Is the Presidio truly nationally significant?" The implied message of this debate negated the value of urban (hence, local) open spaces as nationally significant. Rather, the legacy of the nature/culture dichotomy in the Park Service was to marginalize any park that does not embody one of these dominant ideals. The designation of urban recreation areas as a third but ambiguous category of "recreation" marginalized these new parks and left them vulnerable to criticism. Thus, NRAs figure even less prominently in the hierarchy because they float ambiguously on the margins, reflecting criteria that have not been well defined or promoted as an important ideal. As a result, many questioned whether there is nature in the city and whether this nature is anything more than of local interest. There is great temptation to deny the importance of urban parks or open spaces. One observer summarized the Presidio's dilemma:

> The GGNRA has a tremendous wealth of nationally significant resources that by any standard would meet the designation of a National Park. There are few places in America where visitors can interpret every single military engagement this country has been involved with. It contains a rich tapestry of military history. In addition, it does have an array of natural resources. *If it weren't within this urban setting, it would be a "National Park," not an urban recreation area.* The problem is it's in the city, and so no one takes them seriously as a national park [my emphasis].[20]

Clearly many could not reconcile the concept of a national park in a city; they could not move beyond archetypal notions of nature or culture and national parks. The assumption seems to be that urban areas are somehow neither—whatever nature or culture exists is unlikely to have any significance outside the local context (read: the term *urban national park* is an oxymoron).

A public project of the magnitude of the Presidio required significant advocacy from the Park Service and from others outside. The act of constituency building has become increasingly vital to national parks, especially given the heightened politicization over parks in Congress. A mobilized constituency may influence the outcome of political debate, negotiation, or compromise and can play an important role in the social creation of parks.

This meant one of the most crucial aspects of the Presidio project was constituency building, and one of the most important constituents was, of course, the city. This is deliciously ironic, since historically, antiurbanism was one of the influences that created the national park movement. In addition, cities have been perceived as having little to do with the creation or definition of national parks. On the one hand, the Presidio's geography, its urban context, proved to be a liability for many in Congress. On the other hand, the Presidio's urban context generated effective and vocal advocates for the Presidio in Congress. As the case of the Presidio illustrates, the city played a significant role in the social/political creation of the Presidio as national park.

While I am wary of reifying the city, I do believe that reification offers a useful perspective in this case for several reasons. First, among the hundreds of individuals and businesses that participated in the Presidio planning and transformation, there was a general consensus in favor of the plan, an unusual occurrence in San Francisco. This consensus was most evident in the fact that the city's major daily newspapers reported widespread endorsement of the plan. Second, the body politic—from the mayor to city supervisors to city planners to local congressional representatives—was instrumental in the role of Presidio advocate. Third, numerous individuals, citizen groups, and nonprofit organizations also worked to ensure the Presidio's transformation. Together, these three forces of advocacy and change can be described as representative of the city.

The planning process was about creating a plan for the Presidio, but it was also about engaging an urban community to feel strongly about the national parks. GGNRA Superintendent Brian O'Neill appraised the effort:

> I feel really pleased with the broader sense of community ownership of the Presidio. In the planning process, we were able to get a broad cross-section of the community involved at a level at which they now feel a strong sense of proprietary ownership. They are prepared to fight to ensure its long-term preservation and to ensure that the plan will be adopted and implemented. The public is united. They won't let anyone run too far afoul of the plan and its Grand Vision.[21]

Those who worried that the Presidio would become a national business park underestimated the strong presence of the city. Many argued that the people of San Francisco would provide the strongest controlling mechanism for what happens at the Presidio. San Franciscans exerted and will continue to exert vocal and well-organized public oversight. Although the Presidio Trust is to be directly accountable to Congress, it also will be accountable to the city and the numerous organizations now committed to ensuring the realization of the Grand Vision. The planning process reinforced that.[22] For all the waiting, for all the countless hours spent lobbying for the Presidio, in spite of congressional resistance and occasions of outright hostility, Presidio advocates remain optimistic about the Presidio's future. And they plan to remain vigilant.

An urban experience of parks is crucial, especially given the heightened politicization of national parks. Most Americans live in urban areas; only a small percentage visit Yellowstone and Yosemite. This makes it a challenge to communicate the value of national parks and to educate the wider public about the challenges facing parks. Urban recreation areas, however, can provide both experience and education. The GGNRA, like many urban recreation areas, offers a variety of field trips for city schoolchildren. Public transportation also makes these parks convenient and inexpensively accessible. Visitors to urban recreation areas can experience a diversity of activities. Hopefully, one aspect of their experience will touch them, and they will remember it. In this way will the meaning of national parks be transmitted.

Robert Chandler, the National Park Service's Presidio general manager, posed an interesting twist on the "thinning the blood" notion. He argued that the popularity of the park system in recent years is in some measure due to the fact that the Park Service now has parks in urban areas: "The political value in having a large, diverse system found in both urban settings as well as the so-called crown jewels has added tremendous strength to the system and built popularity to the point where it is very difficult to take on the parks [in Congress]." [23] Because urban parks are more accessible, it is possible that they are building a constituency of people who do not have the opportunity to visit the more isolated national parks. If, in fact, the National Park Service only included national parks of the wilderness sort, the so-called national parks would be not national but regional. Most are located in the West; most are nature parks. These parks would be politically vulnerable in Congress. It is possible—in spite of all the disparaging of urban parks as thinning the blood of the park system—that the nontraditional or urban parks—the GGNRA, the Statue of Liberty, the Gateways—have been instrumental in building a broad base of political support in Congress for all of the national parks. Rather than the issue of worthiness, urban parks perhaps have contributed just the opposite: they have helped educate Americans about the importance of national parks, and they have helped define the national park ideal. In fact, since urban parks expand public access to the national parks, it can be argued that their inclusion in the system was a victory for social justice. It could be argued that urban parks are equally as important as traditional parks in constituency building. These parks may play a more crucial role in the future should Congress pursue a hostile agenda vis-à-vis the national park system.

Nature, Culture, the City, and the Presidio Park

The Presidio's biography is at the heart of this book. But equally important are the broader themes of which the Presidio story is emblematic. One such theme explored the recent changes to the defense structure and the ways in which restructuring and demilitarization can affect the local places; hence the challenge of the Presidio's conversion. A second theme is the exploration of the ways in which people imagine the national park ideal, and the relationship of national parks to broader communities, hence the struggle for the Presidio as a national park.

There were two related dualisms in-between which the Presidio floated ambiguously: nature/culture and wilderness/cities.

Nature and Culture

The first dualism, the opposition of nature and culture, has been well documented and is the subject of a growing body of literature that explores environmental philosophy using social theory, what some commentators have termed the "social construction of nature."[24] This literature reexamines critiques about ideology, discourse, representation, and power. Broadly speaking, social theory overturns many of our assumptions about the modern world by questioning underlying narratives about knowledge. Environmental geographers have recently begun to incorporate these critiques in assessing how we have socially produced place and constructed nature.[25] This literature challenges the assumption that nature is merely a physical entity that can be studied and understood only by natural scientists. Environmental geographers challenge the categorization of nature in scientific terms, arguing that nature is also the product of our social institutions and that how we perceive the natural world reflects issues of equity and distribution, reveals power relationships, and says something about how our culture determines environmental values.

This analysis of nature has arisen from a critical examination of the modernistic paradigm.[26] Generally speaking, the modernistic paradigm refers to a larger worldview or way of looking at the world. Although some disagree about the exact beginning of modernity and the ascendance of the modernistic paradigm, most agree that the roots of the paradigm extend back to the Enlightenment and the Scientific Revolution. One of the primary quests of the Scientific Revolution was to uncover nature's secrets—from the position of Earth in the solar system to the composition of a cell under the microscope. Critics contend that the modernistic paradigm, in an attempt to understand and define the world, set forth certain values that have endured in Western society. They argue this worldview has divided and polarized the world into categories: black/white, male/female, fact/value, science/religion, same/other, nature/culture.[27]

Critics argue that these dualisms are contingent upon the other; the definition of *nature,* for example, relies on the existence of an opposite (*culture*) for meaning.[28] Thus nature was defined as everything but people and their society (or culture)—the nonhuman region. Culture was defined as the product of humanity; it was divorced from nature. Once this myth of nature was established, it was passed on for several generations. Eventually people came to assume that nature was not a product of human imagination but a self-evident "truth."[29]

However, the way in which we think about nature and culture cannot be divorced from the social or intellectual dimensions from where it originated.[30] In fact, critics have discussed the dualism of nature/culture with its inherent paradoxes and contradictions. On the one hand, we seek to understand nature as an object—something "out there" to be studied under a microscope. On the other hand, in pursuing knowledge and experiences in nature, we understand that all life is interrelated.[31] Nothing exists independently, not nature and not humans. There is a place for humanity in nature; nature is, in part, the creation of human imagination of that which is not human.

The social construction of nature literature is useful for understanding the role of culture behind many of our assumed truths about the natural world.[32] Yet the temptation of the modernistic paradigm to promote these dualisms remains powerful. Consider a recent example, the Wilderness Act of 1964, which distinguished between "nature" and "culture": "A wilderness, in contrast with those areas where man and his works dominate the landscape, is hereby recognized as an area where the earth and its community of life are untrammeled by man, where man himself is a visitor who does not remain."

The nature/culture dualism does not really comprise opposites; we just assume that it does. A world divided into opposites is not necessary or even possible. The two parts of a dichotomy are not neatly severed, for they rely on each other for meaning and often their meaning shifts over time. In reality, the world is not either black or white; rather it includes numerous shades of gray and a wonderful spectrum of colors. Dividing the world (or the national parks) into polar opposites erodes the values within the spectrum. The position of the NRAs, for example, illustrates the limitations of these lingering dichotomies as well as the tendency of established "categories" to marginalize entities that reflect a multitude of attributes or fail to meet the national park ideal. Dualisms make no room for contradictions. They make no acknowledgment of the existence of a continuum in which various entities can be valued for their own intrinsic worth. That is one lesson revealed in the Presidio story.

In addition, the valuation of opposites such as nature and culture has shifted over time. Consider the remarkable plasticity and relativity of human values evident in the changing perception of nature in U.S. history.[33] When Europeans first began to settle the North American continent, they reexperienced in America their old, insecure relationship with wilderness.[34] Nature, or the wilderness, was terrifying to them. The colonists perceived the wilderness as physically threatening. On a symbolic level, it represented chaos, immorality. They embarked on a cultural project to subjugate the wilderness, to "civilize" it. Notions of progress meant controlling or eliminating the wilderness. Alexis de Tocqueville wrote in 1835 that Americans rarely thought about the wilderness: "Their eyes are fixed upon another site . . . the march across the wilds, draining swamps, turning the course of rivers, peopling solitudes and subduing nature."[35] The frontier was to be conquered and settled. The respite from the wilds was the city. Cities marked progress and the creation of a civil society. They were places of order and civilization. Cities, therefore, represented the opposite of wilderness.

These perceptions of nature/culture did not last long. By the mid-to-late 1800s, many Americans felt more secure in their ability to control and dominate the wilderness. The chaos in nature had been eliminated by the ax, the plow, the wagon, and the railroad; the frontier was nearly settled. At the same time, urbanization was accelerating rapidly. In the cities, a flood of immigrants, crowded tenements, and intensified industrialization helped to create an artificial but very powerful nostalgia for the outdoors and nature. A new myth about nature was conceived, ironically, in urban terms.[36] Nature became representative of divine transcendence, the sublime, and the wilderness or outdoors became associated with spiritual restoration from crowded and polluted city living. Cities were now perceived as places of chaos and amorality, places to fear. Nature and culture remained opposites, but their meanings had shifted from one to the other. The transposing of meanings attached to the concept of nature/culture thus emerged from historically specific intellectual, social, economic, and political conditions.

The late-nineteenth-century construction of nature as the sublime has persisted. For many people today, nature still represents divine transcendence, and they assume that nature is static, unspoiled, and untrammeled by human activity.[37]

Considering nature as partly a social creation uncovers the instability of the nature/culture dualism. It alerts us to the paradoxes and contradictions of seeing the world as a series of opposites that can be placed in rigid categories. It also alerts us to the consequences of categorization. The literature on this topic is growing but is often complex and obscure. Many works rely on arguments in the abstract; there is little evaluation or exploration of how these dualisms have been practiced and the consequences of such perceptions. A comprehensive discussion of the philosophical tenets of the social construction of nature is beyond the scope of this work. Rather, I believe it important to introduce the argument that nature is partly a social creation because the story of the Presidio reveals that our perceptions of nature, wilderness, and parks do affect the way in which we manage, maintain, or envision nature and national parks. In other words, our conceptions of what is nature, what is culture, and what is a national park do have concrete policy implications. This project has looked critically at the consequences of these persistent dualisms—nature/culture and city/park—by exploring the ways in which these dualisms are, in fact, ephemeral (in the sense that the meaning of these terms has changed and the context in which we perceive these concepts has also changed). The chronicle of the Presidio is a concrete illustration of these themes.

The Social Creation of National Parks

There exists, perhaps, no better example of the artificiality of the nature/culture dualism than the national parks. For many Americans, the national parks—especially Yellowstone, Yosemite, the Grand Canyon, and the Alaskan Wilderness—represent the archetype of nature. National parks are places that many consider profound, inspiring, and magical. They represent spiritual restoration, untainted wilderness, and a commitment to protect and preserve these attributes of nature for future generations.

Those who have studied and written about national parks and the national park ideal do not find the argument that nature is socially constructed surprising. They have long observed that the creation of national parks has been the result of various social movements that became political movements powerful enough to establish federal legislation.

The history of national parks is a history of society's changing attitudes and perceptions about nature and about the role of parks. The creation of the first national parks, Yellowstone and Yosemite, and of every subsequent national park occurred as the end product of a social movement that set in motion a series of cultural events. In every instance, a place captured someone's imagination as a place embodying some important element in the American story—a natural wonder, a site of events important in American history. As more people visited and experienced this place, they came to believe that it was worthy of protection and preservation, and they mobilized. Perhaps they wrote a proposal or drafted legislation documenting the significance of this place. The coalition then secured the sponsorship of a senator or representative who agreed to introduce the legislation to the Congress. Congress voted to approve the legislation, thereby designating this particular place a national park.

National parks, unlike other types of parks, are created through federal legislation. Their designation takes place in the realm of national politics. In order for the legislation to pass, there must be in each house of Congress a majority of members willing to vote to protect this place (their motives may vary: they may truly believe this place worthy of protection; or they may vote in favor of the bill as a favor to another member in return for past favors or future ones). One by one, these steps lead to the creation of a national park. For each national park, there is extensive debate, negotiation, and compromise. Every unit in the park system is therefore an expression of American political culture at a particular time in U.S. history.[38] Each park was at some point a place determined to be nationally significant.

National parks thus exist because of an intricate arrangement of political, social, intellectual, and sentimental relationships. They are as much about statesmanship, philanthropy, and cultural values as they are about ecology. Therefore, national parks are social/political constructions. They reflect a definition (or idealization) of nature, and their creation is reflective of an intent to preserve and protect this ideal for future generations. The creation of national parks and the national park ideal are cultural expressions. They are a time capsule that represents certain values that a society deems important at a particular time in U.S. history.[39] This means the current national park system of approximately 370 parks is a result of more than a hundred years of history involving numerous political decisions and changing cultural contexts.

The social/political creation of national parks has its contradictions and dilemmas. Since national parks are social creations, they have been vulnerable to revision or rejection. There can be no permanence in something socially created (as the changing perception of the wilderness attests). However, the original legislation declared that a national park was to be "inalienable for all time." On the one hand, once a park is created, it should not be decommissioned. On the other hand, the history of national parks is characterized by changing criteria and standards. This is reflected in the addition of new units; it is also evident in the more than sixty parks that have been removed from the jurisdiction of the Park Service.[40] Ideological differences about park standards and about the permanence of the park system are reflected in a current intellectual debate that is increasingly hostile.[41] This ideological debate about standards was evident in the ways in which people perceived the Presidio as part of the GGNRA, an urban national park, and a recreation area.

Nature, Culture, and the Park Service

The national parks also illustrate the contradictions and the consequences associated with insisting on maintaining the nature/culture dualism. For many decades, the Park Service presumed a relatively straightforward dichotomy between two broad categories of parks: natural area parks (nature) and historic/cultural areas and sites (culture).[42] Out of this dual classification arose a hierarchy biased toward the nature parks. This straightforward dichotomy was challenged with the addition of the recreation areas in the 1960s and 1970s. The recreation areas became the third broad category in the park system, but they remained marginal, fitting nowhere within the established hierarchy (and hence relegated to the bottom).

Nature parks are places characterized by quasi-aesthetic romantic notions of the sublime and are considered pristine, as if untouched by culture. Although many believe these parks represent pristine nature or places untrammeled by humans, many studies have shown that the pristine wilderness is only an ideal.[43] Human impact on the land is both inevitable and ubiquitous. Humans have modified or changed the landscape for hundreds and thousands of years, through activities such as clearing and deforestation, setting fires, irrigating, and hunting. Native Americans may have contributed to the mass extinction of some species; European settlers introduced new plant and animal species, thereby altering ecosystem relationships.[44] Much of what we consider pristine or the ideal version of nature is, in fact, a landscape affected to varying degrees by human activity. There is a significant cultural imprint in these nature places. Still, many idealize these nature parks.

Culture parks, in contrast, may be important places, but they do not embody the predominate national park ideal. Although the ultimate ideal of national parks may be the nature parks, culture parks have a long but often overlooked role as important elements in the history of national parks.

The third category, recreation area, has been defined as distinct from traditional park activities such as hiking, enjoying solitude, and reflection. Recreation has come to be defined as undemanding and predictable (adventure, on the other hand, is defined as dealing with the uncertainty of nature). As a result, NRAs have been called playgrounds and open-air gymnasiums, pejoratives that imply a lesser importance than those activities found in the nature parks. Nor are national recreation areas associated with significant contributions to the American identity in the way that a clearly defined culture park may be. I believe the reason for this is that many of the recreation areas are located in urban areas, which leaves these units vulnerable to a lingering antiurban bias underlying the national park ideal. It has also meant that many of these recreation areas must justify their worth as more than locally significant.

These categories within the national parks have had both ideological and policy significance. The perception that nature parks represent the park ideal has created a hierarchy of parks. Some parks are considered more worthy than others. This has led to recurrent arguments that nature parks, as the ideal, deserve more money, more personnel, and more sentiment. Culture parks, in contrast, have often been treated by both the Park Service and Congress as poor stepchildren. More than once individuals or organizations have recommended removing the culture parks from the park system and have suggested that only the nature parks are true national parks. Recreation areas have faced even more hostility and challenge. The legacy of this enduring tension or dualism between nature and culture parks has had an impact on all the national parks, fragmenting the system into categories vague in nomenclature but powerful in symbolism. As Chapter 8 contended, this legacy of nature/culture in the national parks and the marginalization of recreation areas may have subtly influenced the political struggle for the Presidio.

The City and the Park

As much as we celebrate our cities, we also condemn them. In the United States, there has existed a vocal antiurban bias.[45] At the same time that some applaud cities as "civilization," metaphors for individualism, social change, and transformation, others lament the city as an "un-natural" place.[46] To the antiurbanist, cities are full of chaos, crime, disease, violence, poverty, and pollution. They are also devoid of nature.

Cities have been defined as the ultimate expressions of culture.[47] Because cities represent culture, nature has come to represent those characteristics opposite of cities. Nature is pure; it is restorative; it is spiritual. Cities have thus stood apart from nature, a reflection of human beings' ambiguous relationship with the natural world.[48] Cities, therefore, had attributes opposite to those found in national parks. The legacy of an antiurban bias, along with the nature/culture dualism, has meant that few place any significance in the role of the city in the larger matrix of nature.

Historians and geographers have shown that cities modify their physical environment: altering watercourses, creating heat islands, or impacting ecosystems through pollution. Less attention has been paid to the ways in which a city can modify the concept of nature.

This project has shown how the role of the city played a vital, if not the primary, role in the protection, preservation, and creation of the Presidio as a national park. The input of city residents during the planning phase and the mobilization of city officials, individuals, and numerous nonprofit organizations in the role of advocate during the political battle were critical to the transformation of the Presidio and its future land use. These efforts can be seen as a process of reconciliation between nature/culture and cities/national parks.[49] A series of events and actions illustrates this process: the mobilization of the nonprofit organization People for the Golden Gate National Recreation Area and its sponsorship of the GGNRA legislation with its inclusion of the Presidio; the action of municipal officials who zoned the Presidio an urban open space; the public's participation in the planning process, which gave direction to the Presidio's objectives as a national park; the role of city officials and community leaders as advocates for the Presidio in Congress. I argue these efforts should be seen as a social justice movement to "green the city." The city, in this case San Francisco, is directly linked to broader ideas about nature and national parks.

Greening the City

Throughout U.S. history there have been social movements to reclaim the nature in cities. The construction of Central Park in New York City and Golden Gate Park in San Francisco in the mid-to-late 1800s, the City Beautiful movement of the early twentieth century, Robert Moses and his plans for extensive parkways and a state park system for New York state, and the creation of greenbelt suburbs are all examples of such movements.

More recently, a movement to reconcile nature/culture in the city has been called "greening the cities."[50] It has been directly influenced by the environmental movement of the past twenty years, and it revolves around issues of land use and land-use transformation.

Around the world, cities are faced with land-use issues for a variety of reasons. Some cities are making the transformation from industrial city to postindustrial city. They confront abandoned and environmentally degraded landscapes: railway yards, warehouses, obsolete factory sites. Civic leaders search for new ways to transform such landscapes into productive space and to reconnect the community to that space. Other cities, like Seattle or Denver or San Francisco, do not have the industrial legacy to the same degree. Land-use transformations in these cities are often characterized by efforts to reclaim or set aside landscapes for their symbolic value. The preservation of historical districts into "living museums" is one example. Re-creating place-image, through marketing and advertising campaigns, is another, as is the effort to improve city parks and set aside more open space. At the same time, however, these efforts must contend with fiscal pressures on the budgets of municipal parks and recreation departments. Nevertheless, the environment or nature is an important element in urban planning and marketing.

Efforts to protect or expand urban open space reflect a new type of cultural expression: a shift in the perception that there is no nature in cities. This shift is a fundamental rethinking of nature/culture, and it has been influenced by the central message of the environmental movement, which stresses the interconnectedness (not the independence) of nature and culture. Environmentalists have had tremendous influence on the political debate about land use. Recent and dramatic changes to the economy and the urban form have led to an emergence of a new environmentalism, which focuses on quality-of-life issues, including urban land use (or reuse) and the demands for open spaces and green places. In the 1970s, environmentalists primarily mobilized around environmental deterioration. In the 1990s, environmentalists are increasingly mobilizing around the reclamation and preservation of urban open space. The creation of urban farms, the planting of trees along street meridians and in parks, and the building of industrial parks are a few of the ways in which urban environmentalists have begun to influence the creation and preservation of open spaces and green places, however large or small. The movement to green the cities needs to be considered a fundamental cultural rediscovery, and asserts that land-use battles or transformations are cultural expressions.[51]

This is certainly true of the Presidio plan, which reflects ideas and objectives similar to many of those found in green city projects. This is due, in part, to a long legacy of active environmentalism in San Francisco. Since the 1960s, the Bay Area has led the way nationally in environmental awareness and action. It is home to national or regional headquarters for many environmental organizations, including the Sierra Club and Greenpeace. In 1986, several environmental organizations began work to develop a broad-based green city program for the San Francisco Bay Area.[52] Environmentalists were drawn to the Presidio project. Harold Gilliam, local author and environmentalist, explained that "the Environmental Revolution, now beginning to be global in scope, had its roots here in the region inside the Golden Gate, and the Presidio could become the premier place for its expression, celebration and extension."[53] Environmentalists rallied around the Presidio as a symbol of a new urban environmental responsibility, an urban national park.

The conversion of the Presidio illustrates one way to green the city, to revalue an urban space, and to reconnect a community with that place. It represents one attempt by one community to give a new role and meaning to an old military base. Efforts to green the city have occurred in other places in North America. Cities like Pittsburgh (a campaign to clean its image as a gritty industrial city has resulted in significant improvements in air quality as well as numerous park improvements), Cleveland (efforts to revitalize the waterfront area have reconnected people to the Cuyahoga River), Boston (committed to increasing funds for improvements in the Boston Common), Baltimore (the multi-million-dollar Inner Harbor development), and Toronto (redevelopment projects on the waterfront have increased open space) illustrate attempts to reconnect people to their environment. In addition, several cities and communities faced with military base closures are hoping that some of the land may be set aside as open space or nature reserves.[54]

Broad public participation in the planning process is an important element in efforts to green the city. This proved crucial to the Presidio's conversion as well. The local community formed a variety of citizen groups to work with the National Park Service. The Sierra Club established a Presidio task force. Various neighborhood groups set up their own Presidio committees. More recently, several individuals and neighborhood groups organized and formed People for the Presidio, a nonprofit organization to oversee the implementation of the Grand Vision.

Of particular importance in the story of the Presidio was the Presidio Council, which proved to be instrumental in conceptualizing the Presidio Trust partnership. GGNPA also played a crucial advocacy role. Together, these nonprofit, grass-roots organizations and others testified numerous times at congressional hearings, spent countless hours writing letters and organizing arguments to counter those made by hostile congressional representatives, and volunteered their time and their enthusiasm. Their participation has been instrumental in elevating the Presidio from merely a lovely local site to a park of national significance. It is very possible the Presidio legislation might have failed if numerous residents of the city of San Francisco had not been so mobilized, committed, and impassioned.

The Presidio plan has been influenced and promoted by various grassroots groups and nonprofit organizations—from neighborhood coalitions to the elite Presidio Council. They have played a crucial role. Their participation challenged the function and purpose of this space and helped to determine its future meaning. Land-use transformations ultimately change the meaning and form of a city. In the case of the Presidio, the city redefined its relationship to the post and asserted concern for the environment by endorsing the Presidio as an environmental thinktank. Ironically, in this land-use transformation, little of the actual landscape, infrastructure, or buildings will change with the implementation of the Presidio's Grand Vision. Rather, its transformation emphasizes a change in purpose or meaning. It is to be a national park dedicated to the exploration and education of environmental and social problems, a mission much different from that of the military post.

Last Thoughts

It is tempting to divide the world into dichotomies such as nature/culture and city/park because they seem to make the world easier to understand. But this temptation clearly has consequences, as the story of the Presidio has revealed.

The Presidio, of course, challenged many with its complexity. It is simultaneously a nature park, a culture park, and a recreation area. Few nature or culture parks contend with the management of eight hundred buildings. The post's nontraditional history, its form, and its urban context made it especially susceptible to suspicion, resentment, and hostility. The legacy of the nature/culture dualism, the marginalization of the recreation areas, and a lingering antiurban park bias were subtle but powerful and tempted several members of Congress to question the Presidio's designation as a national park.

The contest to realize the Grand Vision and to authorize the Presidio Trust is only one of the more recent struggles to define the role and meaning of national parks. If the Presidio's Grand Vision and the resulting political struggle are any indication, dividing the world into nature/culture, city/park, has policy implications not only for the Presidio but for all national parks. Rather than dividing and classifying the parks into certain categories, which in turn may be internally divided into hierarchies, we should imagine the national parks as a diverse collection of places situated along a continuum of natural and cultural and recreational attributes that reflect America's long and varied history. Sadly, it appears we are far from achieving this more inclusive appreciation of the different parks.

Some observers believe the Presidio represents the prototype of contradictions and challenges that future national parks will face. America's wilderness is, for the most part, protected. This means that future additions to the national park system are likely to be more complex in character: partially developed, mixed-use sites, maybe urban in location, and perhaps environmentally contaminated to some degree. These realities are at odds with predominant park ideals and might generate the same type of heated debate and controversy evidenced in the struggle for the Presidio.

The biography of the Presidio—its history as an Army post, its relationship with the city of San Francisco, the recent conversion planning process, and the struggle to win legislation for the Presidio Trust—has engaged many in a debate that sought to rethink and reconcile nature/culture and city/park. It challenges us to reconsider our biases and assumptions, our tendencies to classify and divide the world. It is not a finished product but an ongoing process.

APPENDIX A

A Note About the Research for This Project

In 1993, I began to conceive of a project that would integrate environmental issues in an urban context. I chose the city of San Francisco as the site for my research for two principal reasons. First, I knew the locale well, having lived in the city for many years. Second, since San Francisco has a well-documented history of environmental activism, I suspected finding a research topic would not prove too difficult. In June 1993, I returned there to spend the summer months on a "site reconnaissance." Professor Donald Kennedy, former president of Stanford University and a good friend, suggested I consider the Presidio project as a potentially fascinating and important case study. He had been involved in the initial planning and gave me access to his archives to familiarize myself with the project. Shortly thereafter another fortunate opportunity arose. William K. Reilly, former administrator of the U.S. Environmental Protection Agency, had relocated to the Bay Area to spend 1993–1994 at Stanford University as a visiting professor. Mr. Reilly was also serving as senior adviser to the Presidio Council. He was in need of a research assistant and, after an interview, offered me the position. I accepted because I knew that for this kind of research it was essential to have a critical entry point that would enable me to access archives, attend meetings, and conduct interviews with more than a dozen key actors, Park Service officials, critics, and other interested parties. From July 1993 through December 1993, I worked closely with Mr. Reilly. His "insider" status in the Presidio Council gave me a unique entry point into the political process of land-use control. My position as his research assistant helped eliminate false starts that might have occurred without this privileged connection. This opportunity also guided my research design. My research relied on qualitative data that I gathered using a combination of three methods: observation, archival work, and interviews.

As an observer, I attended numerous planning and strategy meetings (between fifteen and twenty-five) with the National Park Service, congressional staff assistants, the local Golden Gate National Park Association (GGNPA) administrators, city planners, philanthropic donors, and the Presidio Council. I use the term *observer* because I was not a member of any of the organizations involved, nor did I actively participate in discussions during meetings. However, my status as Mr. Reilly's assistant allowed me to ask the key actors questions in later interviews in order to clarify my observations. I found these meetings invaluable to the research because they revealed an interesting dynamic among the Presidio Council (a citizen group), GGNPA, Park Service officials, congressional members, and outside consultants. In particular, many members of the Presidio Council and staff of GGNPA were frustrated by the lack of planning expertise and financial acumen they judged to exist among Park Service staff. Their perspectives gave me an alternative interpretation of the planning process and the political struggles in Congress. In addition, many of the GGNPA and Presidio Council staff offered an experience of and insight into urban planning and interagency politics that differed from those of newspaper journalists or federal employees entrenched in bureaucratic procedure (such as Park Service officials or members of Congress). This opportunity helped me set the context for understanding the political process that underlies the Presidio plan.

The compilation and analysis of various archival materials were also crucial to the project. Over a three-year period, I collected and analyzed numerous documents, including National Park Service planning documents, drafts of the Presidio plan, maps, historical landmark surveys, public outreach material, congressional records and testimony, letters, and miscellaneous memoranda. Many of these are public documents available in government document collections in many university libraries, or they can be obtained by contacting the appropriate Senate and House subcommittees. In addition to the public documents, I gathered a collection of newspaper articles about the project from major newspapers, including the *San Francisco Chronicle*, the *New York Times*, the *Los Angeles Times*, and the *Wall Street Journal*. I used nearly five hundred newspaper and magazine articles relating to the Presidio and to general base closure transitions. I judged these sources to be important for understanding the way in which the community was apprised of the planning and legislative process. Some articles, such as letters to the editor or editorials, introduced community concerns or criticisms of the Presidio plan. They provided a juxtaposition to the "official" Park Service commentary and public outreach material. I did not use newspaper articles for their analysis of the issue; rather, I used these sources as a starting point for an analysis that relied on other documentation and interview confirmations.

Whereas Chapters 5 to 9 are the result of the analysis of primary interviews and primary and secondary documents, Chapters 2, 3, and 4 (the historical geography) are based predominantly on secondary literature. This use of secondary literature occurred for two reasons. First, many of the archival documents of the Presidio were unavailable during my fieldwork in San Francisco, because the archives were being moved to another site and recataloged. Second, there were two richly detailed and wonderfully comprehensive histories of the Presidio. The first work, written by John

Phillip Langelier and Daniel B. Rosen, *Historic Resource Study: El Presidio de San Francisco, a History Under Spain and Mexico, 1776–1846* (1992), provided an excellent analysis and summary of the archives then unavailable to me. The second work, Erwin Thompson and Sally Woodbridge's *Special History Study: Presidio of San Francisco, an Outline of Its Evolution as a U.S. Army Post, 1847–1990* (1992), was equally excellent.[1] Both of these publications provided a comprehensive history of the Presidio. To my knowledge, they are the only resources that document a history of the Presidio in its entirety. In Chapters 2, 3, and 4, I rely considerably on these two invaluable sources.

My historical geography of the Presidio, however, does add a new and needed contribution to the literature on the Presidio: documentation of the interdependent relationship between the city of San Francisco and the Presidio itself. This relationship was not central to either of the two historical studies of the Presidio, nor has it been the subject of any historical study of the city. Indeed, a survey of some of the most important and influential studies of the history of San Francisco reveals that few integrate the Presidio into the city's history, thus ignoring its importance to and influence in the urban fabric.[2] One exception is Mel Scott's *The San Francisco Bay Area: A Metropolis in Perspective* (1985).[3] In his account of the early years of settling the Bay Area, Scott provides a rich description of the role and importance of the Presidio. But even in his book, some 500 pages long, the Presidio's history and its connection to the city merit only a relatively brief description (less than 15 pages in total). Simply put, the interdependent relationship between the Presidio and San Francisco is a subject that has not been explored, yet as Chapters 2, 3, and 4 show, it is a rich and important interdependence.

Although the historical geography relies on secondary material, I believe I used these various sources to construct an original interpretation of the post's historical geography. I would thus defend this section's methodology on the grounds that while not original research, it is an original application of scholarship and is a vital aspect of the entire story.

I was also fortunate in obtaining documents not available to the public. Mr. Reilly was supportive of my research and insisted that I copy any documents, memos, and miscellaneous information from his extensive archive collection. Other key sources of information were the Presidio Council and GGNPA. These organizations granted me access to their files and gave me copies of planning material, GGNPA public outreach material, and government relations analysis (some of these documents are not available to the public). From the National Park Service Presidio Planning Center, as well as the Presidio archives, I obtained numerous Park Service documents: draft plans, historical studies, architectural surveys, and public outreach publications (many of these publications are available to the public and may be found in government document collections, as well as at the GGNRA Park Archives and Record Center). As a result of my access to these nontraditional sources, I obtained many key and useful documents that aided in my analysis and helped to confirm or triangulate with other information.

My position as Mr. Reilly's assistant also allowed me to meet and interview many of the key actors (for a complete listing of these interviews, see the References). These interviews formed an integral part of the information-gathering process. They clarified conflicting information and revealed unique angles to the story. I was also able to reinterview several of these individuals through e-mail.

I decided to use a blend of structured and unstructured interviews. I began each interview with a common set of questions. The interviews covered various topics including the interviewee's perceptions of the planning process, the final plan document, the battle over the Presidio Trust legislation, Congress as a critical actor, and the weight given to the opinions of local community interests; their interpretations of the interaction between federal agencies, the formation of public-private partnerships, and the support for or hostility to the Presidio within the Park Service; and their predictions for the future of the Presidio. In conjunction with these common questions, I used open-ended questions that were specifically directed toward an individual's or organization's role in the Presidio project. These interviews proved valuable and insightful. They also confirmed my understanding of the general timeline of events and the positions of various organizations. They also gave me an opportunity to gauge reaction to several of my own arguments. Several of the individuals I interviewed told me my questions and observations were thoughtful and probing and asked me to send them a transcript of the interview so they could refer to their responses in future interviews. The information and comments gleaned from these interviews are identified in the Notes. A few individuals requested anonymity because they believed what they said could be perceived as criticism of their employer, and I have honored their request.

These diverse research methods created a flexible but thorough research design that enabled me to develop an interpretive narrative. This project was a wonderful opportunity, perhaps one that is both place and time specific. It also allowed me, I believe, a unique perspective, one unlikely to be easily replicated.

Notes

1. John Phillip Langelier and Daniel B. Rosen, *Historic Resource Study: El Presidio de San Francisco, a History Under Spain and Mexico, 1776–1846* (Denver: U.S. Department of the Interior, National Park Service, 1992); Erwin Thompson and Sally Woodbridge, *Special History Study: Presidio of San Francisco, an Outline of Its Evolution as a U.S. Army Post, 1847–1990* (Denver: U.S. Department of the Interior, National Park Service, 1992).

2. The following works are considered by many to be important contributions to understanding the history of San Francisco, but few of them examine or integrate the Presidio as a component in the history of the city of San Francisco: Herbert Bolten, *Outpost of Empire: The Story of the Founding of San Francisco* (New York: Knopf, 1931); Robert Cleland, *California in Our Time (1900–1940)* (New York: Knopf, 1947); Robert Cleland, *From Wilderness to Empire: A History of California, 1542–1900* (New York: Knopf, 1944); Edgar Kahn, *Cable Car Days in San Francisco*

(Stanford, Calif.: Stanford University Press, 1944); Peter Decker, *Fortunes and Failures: White Collar Mobility in Nineteenth Century San Francisco* (Cambridge, Mass.: Harvard University Press, 1978); Gertrude Atherton, *Golden Gate Country* (New York: Duell, Sloan & Pearce, 1945); Felix Riesenberg, *Golden Gate: The Story of San Francisco Harbor* (New York: Knopf, 1940); Terrence McDonald, *The Parameters of Urban Fiscal Policy: Socio-Economic Change, Political Culture, and Fiscal Policy in San Francisco, 1860–1906* (Berkeley: University of California Press, 1986); Douglas Daniels, *Pioneer Urbanites: A Social and Cultural History of Black San Francisco* (Philadelphia: Temple University Press, 1980); Phillip Ethington, *The Public City: The Political Construction of Urban Life in San Francisco, 1850–1900* (New York: Cambridge University Press, 1994); Roger Lotchin, *San Francisco, 1846–1856: From Hamlet to City* (New York: Oxford University Press, 1974); Oscar Lewis, *This Was San Francisco: Being First-Hand Accounts of the Evolution of One of America's Favorite Cities* (New York: McKay, 1962); Catherine Phillips, *Through the Golden Gate, San Francisco, 1769–1937* (San Francisco: Suttonhouse, 1938). These citations are not repeated in the References at the back of the book.

3. Mel Scott, *The San Francisco Bay Area: A Metropolis in Perspective*, 2nd ed. (Berkeley: University of California Press, 1985).

APPENDIX B

PRESIDIO COUNCIL MEMBERS, 1993

Joan Abrahamson
President
The Jefferson Institute

Carl Anthony
President, Earth Island Institute
Director, Urban Habitat Program

Edward J. Blakely
Professor, Department of City and
Regional Planning
UC Berkeley

Rodger J. Boyd
Executive Director
Division of Economic Development
Navajo Nation

M. J. Brodie
Executive Director
Pennsylvania Avenue Development
Corporation

Dr. Noel J. Brown
Director
Regional Office for North America
U.N. Environment Program

John Bryson
CEO
Southern Cal Edison

Adele Chatfield-Taylor
President
The American Academy in Rome

Richard Clarke
CEO
Pacific Gas & Electric

Francis Ford Coppola
President
American Zoetrope

Robert K. Dawson
Vice Chairman
Cassidy & Associates

Roy Eisenhardt (Vice Chair)
Director
California Academy of Sciences

Patrick Foley
Chairman and President
DHL Corporation

Tully M. Friedman
Hellman & Friedman

Herman E. Gallegos
Chair
Gallegos Institutional
Investors Corporation

John W. Gardner
Haas Professor of Public Service
Stanford University
Graduate School of Business

Jewelle Taylor Gibbs
Professor, School of Social Welfare
UC Berkeley

William Graves
Editor
National Geographic

Walter A. Hass, Jr.[1]
Honorary Chair
Levi Strauss & Co.

James Harvey (Chair)[1]
Chair
Transamerica Corporation

Antonia Hernandez
President and General Council
Mexican American
Legal Defense/Educational Fund

Ira Michael Heyman
Chancellor
UC Berkeley

Roger Heyns
Retired President
William & Flora Hewlett Foundation

Roger Kennedy[2]
Director
National Park Service

Maya Lin
Artist/Architect
(Designer of the Vietnam Memorial)

James Miscoll
Former Vice Chair
Bank of America

Gyo Obata
Chair
Hellmuth, Obata & Kassabaum, Inc.

Laurie Olin
Principal
Hanna/Olin, Ltd.

Toby Rosenblatt (Ex officio)
Chair
GGNPA

Ellen Ramsey Sanger[3]
Executive Director
Coro Foundation

John C. Sawhill
President and CEO
The Nature Conservancy

Dr. Lucy Shapiro
Chair, Department of Developmental Biology
Stanford University School of Medicine

---

[1] Also served as board members of the National Park Foundation; Walter Haas, Jr., served as Director of the National Park Foundation.

[2] When Roger Kennedy was asked to join the Presidio Council, he was Director of the National Museum of American History for the Smithsonian Institution.

[3] The Coro Foundation is an educational institute that conducts public service internships and leadership training projects.

Mimi Silbert [4]
Executive Director
The Delancey Foundation

Virginia Smith
President Emerita
Vassar College

Dr. Bruce Spivey, M.D.
President/CEO
Northwest Healthcare System

Richard A. Trudell
Executive Director
AILTP/American Indian Resources Institute

Robin Winks [5]
Randolph W. Townsend, Jr.,
Professor of History
Yale University

---

[4] The Delancey Foundation is a residential treatment center for felons and substance abusers.

[5] Robin Winks previously served as chair of the U.S. Park System Advisory Board.

APPENDIX C

NATIONAL PARK SERVICE PRESIDIO PLANNING AND TRANSITION TEAM MEMBERS

THE PRESIDIO PLANNING TEAM

Don Neubacher
Supervisory Planner and Team Captain
NPS Denver Service Center*

Roger Kelley Brown
Supervisory Planner
NPS Denver Service Center*

Dennis Piper
Landscape Architect
NPS Denver Service Center*

Carey Feierabend
Historical Architect
NPS Denver Service Center*

Lauren McKean
Community Planner
NPS Denver Service Center*

Aida Parkinson
Natural Resource Specialist
NPS Denver Service Center*

Christopher Schillizzi
Interpretive Specialist
NPS Harpers Ferry Center

Kate Nichol
Public Involvement Specialist
NPS Denver Service Center*

Alison Kendall
Presidio Planning Coordinator
City and County of San Francisco

Kristin Baron
Architectural Technician
NPS Denver Service Center*

Kay Roush
Writer/Editor
NPS Denver Service Center*

Travis Culwell
Architectural Technician
NPS GGNRA

Larry Gill
Assistant
NPS Denver Service Center*

Mai-Liis Bartling
Project Coordinator
NPS GGNRA

Diane Nicholson
Park Curator
NPS GGNRA

Tom Gavin
Regional Forest / Range Conservationist
NPS Denver Service Center*

Terri Thomas
Natural Resource Specialist
NPS GGNRA

The Presidio Transition Team

Mike Savidge
Team Captain

Alex Macievich
Environmental Engineer

John Amos
Architectural Technician

Virginia Phillips
Real Estate Specialist

David Behler
Program Analyst

Gabriele Rennie
Secretary

Lou Ferrari
Space Manager

Chuck Swanson
Civil Engineer

Ron Golerm
Program Analyst

Robert Wallace
Historical Architect

Steve Kasierski
Financial Analyst

---

*Relocated from the Denver Service Center to the Presidio for the duration of the planning process.

Notes

Notes to Chapter 1

1. National Park Service, *Creating a Park for the 21st Century: From Military Post to National Park. Draft General Management Plan Amendment, Presidio of San Francisco,* NPS D-148 (San Francisco: GGNRA, October 1993).

2. Some historical geographers would argue that all of these constitute current historical geography (Anne Mosher, professor of geography, Syracuse University, personal communication, 1996). I, however, would make a distinction between an analysis and documentation of the past (historical geography) and an analysis of the current (in this project it involves a combination of political and cultural geography).

3. The introduction to this book has been shaped for a more general interest audience. For those with an interest in my theoretical elaboration for this project, I direct you to Chapter 9, Appendix A (a note about my research methodology), and several of my publications. Together, these begin to articulate an emerging theory of the relationship between society and space, nature and culture, and environmental ideology and political action. It is my hope that this and future works will unravel connections that have not been previously considered.

My previously published works that consider the above themes from various angles using several case studies are: Lisa Benton, "The Greening of World Trade?: Reconciling Economics and the Environment in the Debate about the North American Free Trade Agreement," *Environment and Planning A* 28, 1996: 2155–2177; Lisa Benton, "Will the Reel/Real Los Angeles Please Stand Up?" *Urban Geography* 16(2), 1995: 144–164; Lisa Benton, "Selling Nature or Selling Out? Exploring the Contradictions of Environmental Merchandise," *Environmental Ethics* Spring 17(1), 1995: 3–22; John Rennie Short, Lisa Benton, Blake Luce, and Judy Walton, "Reconstruction of a Postindustrial City," *Annals of the American Geographers* 83, 1993: 207–224.

NOTES TO CHAPTER 2

1. Two excellent sources on the Ohlone culture are Malcolm Margolin, *The Ohlone Way: Indian Life in the San Francisco–Monterey Bay Area* (Berkeley, Calif.: Heyday Books, 1978); and John Bean Lowell, ed., *The Ohlone Past and Present: Native Americans of the San Francisco Bay Region* (Menlo Park, Calif.: Ballena Press, 1994).

2. Margolin, *The Ohlone Way*, 1.

3. Ibid., 40.

4. Ibid., 59; see also Alfred Kroeber, *Handbook of the Indians of California* (Washington, D.C.: GPO, 1925).

5. Kroeber, ibid.

6. The term "Alta California" refers to the present-day state of California, from San Diego northward. Although Spain had not established settlements in Alta California, it was firmly established in "Baja California," which today remains part of Mexico.

7. John Phillip Langelier and Daniel B. Rosen, *Historic Resource Study: El Presidio de San Francisco, a History Under Spain and Mexico, 1776–1846* (Denver: U.S. Department of the Interior, National Park Service, 1992: 2).

8. Fray Francisco Paloú, *Life and Apostolic Labors of the Venerable Father Junipero Serra* (Pasadena, Calif.: G. W. James, 1913). See also Margolin, *The Ohlone Way*, 159.

9. Langelier and Rosen, *Historic Resource Study*, 2.

10. Mel Scott, *The San Francisco Bay Area: A Metropolis in Perspective*, 2nd ed. (Berkeley: University of California Press, 1985: 6).

11. Langelier and Rosen, *Historic Resource Study*, 6.

12. Ibid., 6.

13. Scott, *The San Francisco Bay Area*, 9.

14. Herbert Eugene Bolton, ed., *Anza's California Expedition*, vol. 4 (Berkeley: University of California Press, 1930: 329–339), and quoted in Langelier and Rosen, *Historic Resource Study*, 9.

15. Bolton, ibid., 340–343, and quoted in Langelier and Rosen, ibid., 9–10.

16. Kevin Starr, *Americans and the California Dream, 1850–1915* (Santa Barbara, Calif.: Peregrine Smith, 1973: 6).

17. David Hornbeck, "Spanish Legacy in the Borderlands," in *The Making of the American Landscape*, ed. Michael Conzen (Boston: Unwin Hyman, 1990: 55).

18. Langelier and Rosen, *Historic Resource Study*, 11.

19. Neal Harlow, *California Conquered: War and Peace in the Pacific, 1846–1850* (Berkeley: University of California Press, 1982: 20).

20. Hornbeck, "Spanish Legacy in the Borderlands," 57.

21. John Berger, *The Franciscan Missions of California* (Garden City, N.Y.: Doubleday, 1941: 41).

22. Langelier and Rosen, *Historic Resource Study*, 12.

23. Herbert Eugene Bolton, ed., *The Historical Memoirs of New California by Fray*

Francisco Palóu, O.F.M., vol. 4 (New York: Russell and Russell, 1926 and 1966: 125–127), and quoted in Langelier and Rosen, *Historic Resource Study*, 12. In addition, Margolin suggests that the Ohlones had two reactions to the Spanish invasion. It is likely that many of them fled into the hills when the Spaniards arrived. But many of those remaining behind were eager to make contact with the men whose skin, eyes, and hair were of colors different from those of any humans they had ever seen. They were keenly interested in trading cloth, glass beads, metal, and mules (Margolin, *The Ohlone Way*, 157–158).

24. Tom Cole, *A Short History of San Francisco* (San Francisco: Don't Call It Frisco Press, 1981: 15).

25. Berger, *The Franciscan Missions of California*; Margolin, *The Ohlone Way*, 163.

26. Margolin, ibid., 159.

27. Langelier and Rosen, *Historic Resource Study*, 17.

28. Ibid., 27.

29. Ibid., 35.

30. Marguerite Eyer Wilbur, ed., *Vancouver in California, 1792–1794: The Original Account of Vancouver*, vol. 1 (Los Angeles: Glen Dawson, 1954: 12–33).

31. Langelier and Rosen, *Historic Resource Study*, 69.

32. Starr, *Americans and the California Dream*, 5.

33. Langelier and Rosen, *Historic Resource Study*, 103.

34. Quoted ibid., 120.

35. Quoted ibid., 119.

36. Starr, *Americans and the California Dream*, 10.

37. Richard Henry Dana, *Two Years Before the Mast* (Roslyn, N.Y.: Classics Club, 1840: 176–177).

38. Starr, *Americans and the California Dream*, 415.

39. Langelier and Rosen, *Historic Resource Study*, 188.

40. Harlow, *California Conquered*, 4.

41. Marguerite Eyer Wilbur, ed., *Duflot De Mofra's Travels on the Pacific Coast*, vol. 1 (Santa Ana, Calif.: Fine Arts Press, 1937: 228–229), and quoted in Langelier and Rosen, *Historic Resource Study*, 120.

42. Quoted in Langelier and Rosen, ibid., 121–123.

43. Ibid., 123.

Notes to Chapter 3

1. John Phillip Langelier and Daniel B. Rosen, *Historic Resource Study: El Presidio de San Francisco, a History Under Spain and Mexico, 1776–1846* (Denver: U.S. Department of the Interior, National Park Service, 1992: 128).

2. Mary Lee Spence and Donald Jackson, eds., *The Expedition of John C. Frémont* (Urbana: University of Illinois Press, 1973: 183).

3. Erwin Thompson and Sally Woodbridge, *Special History Study: Presidio of San Francisco, an Outline of Its Evolution as a U.S. Army Post, 1847–1990* (Denver: U.S. Department of the Interior, National Park Service, 1992: 7).

4. The name Yerba Buena, or "Good Herb," came from an aromatic vine that grew in abundance in the area. See John Berger, *The Franciscan Missions of California* (Garden City, N.Y.: Doubleday, 1947: 344).

5. Tom Cole, *A Short History of San Francisco* (San Francisco: Don't Call It Frisco Press, 1981: 26).

6. Kevin Starr, *Americans and the California Dream, 1850–1915* (Santa Barbara, Calif.: Peregrine Smith, 1973: 51).

7. National Park Service, *Presidio of San Francisco National Historic Landmark District: Historic American Buildings Survey Report* (San Francisco: U.S. Department of the Interior and U.S. Department of Defense, 1985: 8).

8. Mel Scott, *The San Franciso Bay Area: A Metropolis in Perspective*, 2nd ed. (Berkeley: University of California Press, 1985: 31).

9. Ibid., 24.

10. National Park Service, *Presidio of San Francisco*, 9.

11. Thompson and Woodbridge, *Special History Study*, 9.

12. National Park Service, *Presidio of San Francisco*, 9.

13. Quoted in Thompson and Woodbridge, *Special History Study*, 13.

14. Ibid., 8.

15. Ibid., 17.

16. Elisha Smith Capron, *History of California* (Boston: publisher unavailable, 1854: 129).

17. Thompson and Woodbridge, *Special History Study*, 2.

18. Ibid., 22.

19. Starr, *Americans and the California Dream*, 122.

20. Thompson and Woodbridge, *Special History Study*, 30.

21. See, for example, Martin Melosi, ed., *Pollution and Reform in American Cities, 1870–1930* (Austin: University of Texas Press, 1980); Martin Melosi, *Garbage in the Cities: Refuse, Reform and the Environment, 1880–1980* (College Station: Texas A&M University Press, 1981).

22. This broad reform movement is often considered the precursor of the reforms of the Progressive era.

23. Roderick Nash, *Wilderness and the American Mind* (New Haven, Conn.: Yale University Press, 1982); Joseph Petulla, *American Environmental History*, 2nd ed. (Columbus, Ohio: Merrill, 1988).

24. Petulla, ibid., 237.

25. See, for example, Melosi, *Pollution and Reform;* Edward K. Muller, "The Americanization of the City," in *The Making of the American Landscape*, ed. Michael Conzen (Boston: Unwin Hyman, 1990: 269–293); David Ward, *Cities and Immigrants:*

A Geography of Change in Nineteenth Century America (New York: Oxford University Press, 1971).

26. Peter Schmitt, *Back to Nature: The Arcadian Myth in Urban America* (Baltimore: Johns Hopkins University Press, 1990).

27. Scott, *The San Francisco Bay Area*, 51.

28. Starr, *Americans and the California Dream*, 123.

29. Scott, *The San Francisco Bay Area*, 53.

30. Quoted in Raymond Clary, *The Making of Golden Gate Park: The Early Years, 1865–1906* (Sausalito, Calif.: California Living Books, 1980: 2).

31. Thompson and Woodbridge, *Special History Study*, 34. The authors do not mention the names of these citizens, nor do they cite the original source of this information.

32. The building of Golden Gate Park is a topic unto itself. For a thorough history of its construction, see Clary, *The Making of Golden Gate Park*.

33. *Daily Alta California*, April 7, 1870.

34. Thompson and Woodbridge, *Special History Study*, 34.

35. Ibid., 36.

36. Ibid., 37.

37. Ibid., 38.

38. Quoted ibid., 38.

39. Construction on military bases around the United States followed whatever architectural style was favored by federal planners. Military bases thus were subject to central planning and followed more universal styles rather than the influence of local architecture. Wilbur Zelinsky addresses the influence of government planning on the look of the American landscape. See Wilbur Zelinsky, "The Imprint of Central Authority," in Conzen, *The Making of the American Landscape*, 335–354.

40. Schmitt, *Back to Nature*, 59.

41. Ibid.

42. Clary, *The Making of Golden Gate Park*, 24.

43. Quoted in Thompson and Woodbridge, *Special History Study*, 53.

44. Zelinsky, "The Imprint of Central Authority."

45. Thompson and Woodbridge, *Special History Study*, 57.

46. Harold Gilliam, *The Natural World of San Francisco* (Garden City, N.Y.: Doubleday, 1967: 82).

47. Ibid., 84.

48. Ibid.

49. Denis Cosgrove and Stephen Daniels, eds., *The Iconography of Landscape: Essays on the Symbolic Representation, Design and Use of Past Environments* (Cambridge: Cambridge University Press, 1988: 43).

50. Gilliam, *The Natural World of San Francisco*, 119.

51. Starr, *Americans and the California Dream*, 138.

52. Ibid.

53. William H. Wilson, *The City Beautiful Movement* (Baltimore: Johns Hopkins University Press, 1990).

54. Ibid.

55. Donald Krueckeberg, *Introduction to Planning History in the United States* (New Brunswick, N.J.: Center for Urban Policy Research, 1983: 5).

56. Quoted in Starr, *Americans and the California Dream*, 290.

57. Wilson, *The City Beautiful Movement*, 17.

58. Starr, *Americans and the California Dream*, 291.

59. Cited in Scott, *The San Francisco Bay Area*, 105.

60. Robert W. Cherny, "City Commercial, City Beautiful, City Practical: The San Francisco Visions of William C. Raboth, John Phelan and Michael M. O'Shaughnessy," *California History* 73(4), 1995: 303.

61. I describe the Presidio as a "cultural landscape" to connote a landscape that is more than just a collection of buildings of various architectural styles. The term refers to changes and "improvements" to the natural environment, such as tree planting, landscaping, laying out roads, and locating various activities. All of these imprints on the landscape are the result of human decision making and actions and thus constitute efforts at creating meaning in a place as well as function.

62. Thompson and Woodbridge, *Special History Study*.

Notes to Chapter 4

1. Tom Cole, *A Short History of San Francisco* (San Francisco: Don't Call It Frisco Press, 1981: 104–105).

2. Erwin Thompson and Sally Woodbridge, *Special History Study: Presidio of San Francisco, an Outline of Its Evolution as a U.S. Army Post, 1847–1990* (Denver: U.S. Department of the Interior, National Park Service, 1992: 78).

3. Quoted in Mel Scott, *The San Francisco Bay Area: A Metropolis in Perspective*, 2nd ed. (Berkeley: University of California Press, 1985: 113).

4. Kevin Starr, *Americans and the California Dream, 1850–1915* (Santa Barbara, Calif.: Peregrine Smith, 1973: 293).

5. Ibid., 412. This interest in Spanish Revival architecture and building was not limited to California or the Southwest; it was part of a broader movement known as the Arts and Crafts movement, which was sweeping through Europe and the United States. Frank Lloyd Wright in the Midwest, Charles Rennie MacKintosh in Scotland, and William Morris in England all championed the new style. Leopold Stickley and his brother John George founded Stickley, Inc. and in 1905 introduced their Mission Oak furniture line. Their workshop was located in Syracuse, New York.

6. National Park Service, *Presidio of San Francisco National Historic Landmark District: Historic American Buildings Survey Report* (San Francisco: U.S. Department of the Interior and U.S. Department of Defense, 1985: 18).

7. Ibid., 17.
8. Quoted in Thompson and Woodbridge, *Special History Study*, 78.
9. National Park Service, *Presidio of San Francisco*, 18.
10. Starr, *Americans and the California Dream*, 298.
11. Cole, *A Short History of San Francisco*, 116.
12. Thompson and Woodbridge, *Special History Study*, 98.
13. John Jakle, "Landscapes Redesigned for the Automobile," in *The Making of the American Landscape*, ed. Michael Conzen (Boston: Unwin Hyman, 1990: 295).
14. Scott, *The San Francisco Bay Area*, 175.
15. National Park Service, *Presidio of San Francisco*, 20.
16. Ibid.
17. Ibid., 21.
18. Thompson and Woodbridge, *Special History Study*, 198.
19. Quoted ibid., 113.
20. Scott, *The San Francisco Bay Area*, 236.
21. Thompson and Woodbridge, *Special History Study*, 116.
22. National Park Service, *Presidio of San Francisco*, 19.
23. Ibid.
24. Cole, *A Short History of San Francisco*, 124–128.
25. Ibid., 129.
26. Scott, *The San Francisco Bay Area*, 244.
27. Ibid., 256.
28. Golden Gate National Park Association, *Presidio Gateways: Views of a National Historic Landmark at San Francisco's Golden Gate* (San Francisco: Chronicle Books, 1994: 56).
29. Ibid., 11–14.
30. Truman may have thought of the Presidio for two reasons: (1) he was in San Francisco at the time of the conference, and (2) he may have understood that the Presidio was no longer the vital defense center it once had been.
31. Thompson and Woodbridge, *Special History Study*, 123.
32. Ibid.
33. National Park Service, *Presidio of San Francisco*, 23.
34. Thompson and Woodbridge, *Special History Study*, 1992.
35. Ibid., 124.
36. Ibid., 125.
37. National Park Service, *Presidio of San Francisco*, 23.
38. B. Brugmann and G. Stetteland, *The Ultimate Highrise* (San Francisco: SF Bay Guardian Books, 1971: 30).
39. Herb Caen, *One Man's San Francisco* (Sausalito, Calif.: Comstock Editions, 1976: 50). Herb Caen began writing about San Francisco in 1938. Over the course of fifty-eight years, his daily column helped to define the city. "Herb Caen really is San Francisco," said Mayor Willie Brown. "He's done more than any human being to en-

dear San Francisco to the world" (quoted in Steve Rubenstein, "Readers Far and Wide Overwhelm Caen with Good Wishes," *San Francisco Chronicle,* May 31, 1996: E1). As testimony to Caen's popularity and position, in May 1996 the city renamed the 3.2-mile waterfront promenade known as the Embarcadero "Herb Caen Way."

40. Scott, *The San Francisco Bay Area,* 274.

41. U.S. House of Representatives, Committee on Interior and Insular Affairs, Subcommittee on National Parks and Recreation, *Hearings on H.R. 9498 to Establish a National Recreation Area in San Francisco and Marin Counties, August 9 and May 11 and 12, 1972* (Washington, D.C.: GPO, 1972: 22).

42. Richard DeLeon, *Left Coast City: Progressive Politics in San Francisco, 1975–1991* (Lawrence: University Press of Kansas, 1992).

43. Ibid., 3.

44. See Alfred Heller, ed., *The California Tomorrow Plan* (Los Altos, Calif.: William Kaufmann, 1972).

45. Scott, *The San Francisco Bay Area,* 281.

46. Ibid., 315.

47. Ibid., 327.

48. U.S. House of Representatives, *Hearings on H.R. 9498,* 399.

49. Alfred Runte, *National Parks: The American Experience,* 2nd ed. (Lincoln: University of Nebraska Press, 1987: 233).

50. Jerry Carroll, "Saviors of the Land," *San Francisco Chronicle,* January 16, 1997: B3.

51. Ibid.

52. U.S. House of Representatives, *Hearings on H.R. 9498,* 14.

53. Supporting organizations included the Presidio Society, Association of Bay Area Governments, California Historical Society, Sierra Club, People for Open Space, National Parks and Conservation Association, San Francisco Chamber of Commerce, League of Women Voters of the Bay Area, Headlands Tomorrow, Save the Bay, San Francisco Tomorrow, Citizens Waterfront Committee of San Francisco, Golden Gate Audubon Society, San Francisco Beautiful, Barristers Club of San Francisco, San Francisco Maritime Museum, National Wildlife Federation, East Bay Citizens for a GGNRA, San Francisco Bicycle Coalition, and California Tomorrow.

54. U.S. House of Representatives, *Hearings on H.R. 9498,* 62.

55. Ibid., 14.

56. Ibid., 21–22.

57. Ibid., 253.

58. This information taken from a PFGGNRA document, ibid., 140–142.

59. The year Congress created the GGNRA it also passed legislation creating the Gateway National Recreation Area in New York and New Jersey. In fact, Burton became a leading crusader for parks. As chair of the House Subcommittee on National Parks and Insular Affairs, Burton was instrumental in winning passage of the National Parks and Recreation Act of 1978. This legislation incorporated a host of

new projects, increased appropriations and acquisitions, made some boundary changes, and authorized several new parks (the Santa Monica Mountains National Recreation Area and the New River Gorge National River in West Virginia). Runte, *National Parks,* concludes this legislation was both impressive and unprecedented.

60. Public Law 92-589.

61. Carroll, "Saviors of the Land," B3.

62. See, for example, H. Duane Hampton, *How the U.S. Cavalry Saved Our National Parks* (Bloomington: Indiana University Press, 1971).

63. U.S. House of Representatives, *Hearings on H.R. 9498,* 26.

64. Vikram Seth, *The Golden Gate, a Novel in Verse* (New York: Vintage Books, 1986: 103–104).

65. Edward Relph, *Place and Placelessness* (London: Plion, 1976: 41).

66. Ibid., 43.

67. Derwent Whittlesey, "Sequent Occupance," *Annals of the Association of American Geographers* 19, 1929: 162–166.

68. U.S. Senate, *Management of the Presidio in San Francisco by the National Park Service, Hearing Before the Subcommittee on Public Lands, National Parks, and Forests of the Committee on Energy and Natural Resources, May 12, 1994* (Washington, D.C.: GPO, 1994: 14).

Notes to Chapter 5

1. Several outspoken critics of the arms race and the military-based economy have quoted this scripture verse. Peace activists too have used it. I include it here because many supporters refer to base closures and defense downsizing as offering not only an economic opportunity but also a way to transform society.

2. The terms *defense drawdown, downsizing,* and *reduction* are interchangeable and are used as such in Department of Defense reports and published documents. So too are *demobilization* and *demilitarization.* These terms indicate reduction in spending and in personnel (both military and civilian) and in some cases the closure of military installations. I will also use these terms interchangeably.

3. Seymour Melman, *The Demilitarized Society: Disarmament and Conversion* (Montreal: Harvest House, 1988: 24).

4. The term *economic conversion* is associated with a defense facility or laboratory that was originally constructed for the purpose of defense research or production and is to be converted to civilian use.

5. U.S. Department of Defense, Defense Conversion Commission, *Adjusting to the Drawdown: Report of the Defense Conversion Commission* (Washington, D.C.: GPO, December 31, 1992: 12).

6. Breandan OhUallachain, "Regional and Technological Implications of the Recent Buildup in American Defense Spending," *Annals of the Association of American Geographers* 77, 1987: 208–223.

7. John E. Lynch, *Local Economic Development After Military Base Closures* (New York: Praeger, 1970: ix); see also U.S. Arms Control and Disarmament Agency, *Economic Impact of Military Base Closings: Adjustments by Communities and Workers*, vol. 1 (Washington, D.C.: GPO, 1970). A defense installation includes any of the following facilities: a military base, camp, post, station, depot, warehouse, homeport, yard. For the purposes of this chapter, I consider major military base closures and alignments. The Defense Department defines a major military base as a base that employs at least three hundred civilians.

8. Lynch, *Local Economic Development*, 7.

9. Ibid., 9.

10. Quoted ibid., 6.

11. Betty Lall and John Tepper Marlin, *Building a Peace Economy: Opportunities and Problems of Post–Cold War Defense Cuts* (Boulder, Colo.: Westview Press, 1992: 7).

12. Melman, *The Demilitarized Society*, 21–22.

13. See, for example, the works Lynch, *Local Economic Development*; Seymour Melman, *Planning for Conversion of Military-Industrial and Military Base Facilities* (Washington, D.C.: U.S. Department of Commerce, Economic Administration, Office of Technical Assistance, 1973); Seymour Melman, "Swords into Ploughshares," *Technology Review* 89(1), 1986: 37–45; Melman, *The Demilitarized Society*; Suzanne Gordon and Dave McFadden, eds., *Economic Conversion: Revitalizing America's Economy* (Cambridge, Mass.: Ballinger, 1984).

14. Melman, *The Demilitarized Society*, 14; Seymour Melman, *The Permanent War Economy* (New York: Simon and Schuster, 1985); Hugh G. Mosely, *The Arms Race: Economic and Social Consequences* (Lexington, Mass.: Heath, 1985).

15. Seymour Melman, "Economic Consequences of the Arms Race: The Second-Rate Economy," *American Economic Review*, May 1988: 57.

16. Peter Hall and Ann Markusen, "The Pentagon and the Gunbelt," in *The Pentagon and the Cities*, ed. Andrew Kirby (Newbury Park, Calif.: Sage, 1992: 53).

17. Ann Markusen, Peter Hall, Scott Campbell, and Sabina Deitrick, *The Rise of the Gunbelt: The Military Remapping of Industrial America* (New York: Oxford University Press, 1991). The term "Gunbelt," like "Sunbelt" and "Rustbelt," accurately describes both the concentration of defense spending and, not unconnectedly, the economically healthier areas in the United States during the late 1970s, and throughout the 1980s. See Larry Sawers and William K. Tabb, eds., *Sunbelt/Snowbelt: Urban Development and Regional Restructuring* (New York: Oxford University Press, 1984).

18. Andrew Kirby, ed., *The Pentagon and the Cities* (Newbury Park, Calif.: Sage, 1992); Edward J. Malecki, "Military Spending and the U.S. Defense Industry: Regional Patterns of Military Contracts and Subcontracts," *Environment and Planning C: Government and Policy* 2, 1984: 31–44.

19. Indeed, this restructuring process has been referred to by the media, as well as by Congress and Defense Department agencies, as "post–Cold War military base closures." See, for example, U.S. Congress, Office of Technology Assessment, *After*

the Cold War: Living with Lower Defense Spending (Washington, D.C.: GPO, February 1992).

20. Ibid.

21. Peter Trubowitz and Brian Roberts, *Bearing the Burden: Distributive Politics and National Society,* Working Paper, John M. Olin Institute for Strategic Studies, Harvard University, 1991: 11.

22. Cited in U.S. House of Representatives, Committee on Banking, Finance, and Urban Affairs, *Economic Conversion: Hearing Before the Subcommittee on Economic Stabilization, June 29, 1988* (Washington, D.C.: GPO, 1988: 41).

23. Ann Markusen and Joel Yudken, *Dismantling the Cold War Economy* (New York: Basic Books, 1992: 38).

24. Ibid., 133.

25. Ibid., 38.

26. U.S. Congress, *Economic Conversion.*

27. Seymour Melman, an economist and long-time scholar of the military-industrial complex, and Robert DeGrasse have written extensively about the impact of a military economy on American economic vigor and the loss of industrial supremacy. See Melman, *The Permanent War Economy* and "Economic Consequences of the Arms Race," and Robert DeGrasse, *Military Expansion, Economic Decline: The Impact of Military Spending on U.S. Economic Performance* (Armonk, N.Y.: M. E. Sharpe, Council on Economic Priorities, 1983). See also the recent work by Paul Kennedy, *The Rise and Fall of the Great Powers: Economic Change and Military Conflict from 1500–2000* (New York: Vintage Books, 1987).

28. Jonathan Feldman, *An Introduction to Economic Conversion: Briefing Paper One* (Washington, D.C.: National Commission for Economic Conversion and Disarmament, May 1988: 3); my emphasis.

29. David Rapp, "Deficit Limits Reagan's Options in 1989 Budget," *Congressional Quarterly,* February 20, 1988: 331.

30. Ibid., 327.

31. Pat Towell, "The Carlucci Budget: Picking Among Priorities," *Congressional Quarterly,* February 27, 1988: 522.

32. Rapp, "Deficit Limits Reagan's Options," 327.

33. Pat Towell, "Tough Challenges Await the 41st President," *Congressional Quarterly,* September 10, 1988: 2491.

34. John Felton and Pat Towell, "Bush Inherits Political Opportunities, Risk," *Congressional Quarterly,* December 17, 1988: 3503.

35. U.S. Congress, *Economic Conversion,* 11–15.

36. Strobe Talbot, "Reciprocity at Last," *Time,* December 18, 1989: 40.

37. "After the Cold War: Do We Need an Army?" cover page, *U.S. News & World Report,* December 11, 1989; "The Russian's Aren't Coming," *Newsweek,* November 27, 1989: 48–50; "Rethinking the Red Menace," *Time,* January 1, 1990: 66–72.

38. Quoted in Elizabeth Palmer, "Commission Comes to Life, Vowing a Fresh Look," *Congressional Quarterly*, April 20, 1991: 994.

39. Elizabeth Palmer and Pat Towell, "Cheney Reveals New Hit List; Members Feel the Pain," *Congressional Quarterly*, April 13, 1991: 931.

40. Ibid., 931.

41. U.S. Department of Defense, Base Closure and Realignment Commission, *Base Closure and Realignment Commission Report, 1995* (Washington, D.C.: GPO, March 1995: E1).

42. Mike Mills, "Members Go on the Offensive to Defend Bases," *Congressional Quarterly*, July 2, 1988: 1817.

43. U.S. Congress, *Economic Conversion*.

44. Quoted in Josh Getlin, "Congress Passes Plan to Cut Bases," *Los Angeles Times*, October 13, 1988: A1.

45. William Taft, "Editorial: When Doves Turn into Hawks," *Washington Times*, October 5, 1987: D1.

46. U.S. House of Representatives, Committee on Armed Services, *Hearing on Base Closure: Military Installations and Facilities Subcommittee Hearing on H.R. 1583 and Military Installations and Facilities Subcommittee and Defense Policy Panel Joint Hearing on H.R. 4481, March 17, 18, 19, and June 8, 1988* (Washington, D.C.: GPO, 1988).

47. Christine Lawrence, "House Modifies, Passes Base-Closing Bill," *Congressional Quarterly*, July 16, 1988: 1976.

48. U.S. Congress, *Hearing on Base Closure*, 12.

49. Ibid.

50. U.S. Congress, *Economic Conversion*, 21.

51. Armey is currently majority leader of the House of Representatives.

52. Neil Brown, "Base-Closing Process Thwarts Parochialism," *Congressional Quarterly*, July 10, 1993: 1842.

53. Pat Towell, "Hill Paves Way for Closing Old Bases," *Congressional Quarterly*, October 15, 1988: 2999.

54. U.S. Department of Defense, Defense Secretary's Commission on Base Realignment and Closure, *Report of the Defense Secretary's Commission* (Washington, D.C.: GPO, December 1988). BRAC drew upon a list of suggestions proposed by Secretary Cheney. The report was made public a few days later, in January 1989. This is why the first round of base closures is often referred to as the 1988–1989 round.

55. Ibid., 370.

56. Reagan signed the BRAC recommendation into law just a week before George Bush was inaugurated.

57. The next chapter discusses in detail the reactions to the Presidio's closure and Boxer's and Pelosi's attempts to exempt the post from the BRAC list.

58. Quoted in Mike Mills, "Base Closings: The Political Pain Is Limited," *Congressional Quarterly*, December 31, 1988: 3628. I find it revealing that Aspin describes

military base closures in terms of "congressional members hit," rather than referring to cities, states, or the local communities or people affected by base closures. This seems to highlight the very political nature of base closures in Congress, which successfully thwarted base closures for so long.

59. Kirby, *The Pentagon and the Cities*, 15.

60. U.S. Congress, House of Representatives, Committee on Armed Services, *Report of the Defense Secretary's Commission on Base Realignment and Closure Hearings Before the Military Installations and Facilities Subcommittee, February 22 and March 1, 1989* (Washington, D.C.: GPO, 1989: 478). This may be the intention. However, one case stands out. The state of Georgia is home to twenty-one military installations including eleven major military bases. Georgia lost no major military bases in any round of base closures; it is the only state with significant installations to emerge from the four rounds of base closures without a single major base closed. I suspect that perhaps one reason may be that Georgia senator Sam Nunn chaired the Senate Armed Services Committee during all rounds of base closure decision making. The "Nunn factor" may indicate that although Congress established a bipartisan commission to remove the "politics" associated with military base closures, it did not remove political considerations completely.

61. Elizabeth Palmer, "Commission Spares Just a Few from Cheney's Hit List," *Congressional Quarterly*, July 6, 1991: 1845–1847.

62. Defense Department figures cited in Pat Towell, "Tough Calls Are Put on Hold; Clinton Keeps Options Open," *Congressional Quarterly*, March 20, 1993: 679.

63. U.S. Department of Defense, *Base Closure and Realignment Commission Report*, 3.

64. Ibid., 141.

65. This may represent only a small portion of the overall defense budget, but as one Washington politician aptly observed, "a million here, a billion there, pretty soon it starts to add up to real money."

Notes to Chapter 6

1. U.S. Department of Defense, Defense Secretary's Commission on Base Realignment and Closure, *Report of the Defense Secretary's Commission* (Washington, D.C.: GPO, December 1988).

2. Although the Presidio was relatively clean for a military installation, environmental hazards on the post included leaking gasoline tanks, asbestos-laced buildings, electrical transformers loaded with PCBs and several dump sites. This collection of hazards was not surprising since military bases are often exempt from state environmental standards and regulations. Also, environmental regulations are relatively recent (since 1970); many military bases have a long history of leaks and pollutants that precede (and hence have been exempted from) recent regulations.

3. See the preceding chapter's discussion about the end of base-by-base bickering brought about by the Base Closure Act.

4. Many retired military personnel worried specifically about the consequences of closing the Letterman Military Hospital. Many of them had settled in the city expressly because of the military-health-care system, and they were concerned that there would be no military medical services within easy access. Retired Army lieutenant colonel Joseph Arrigo echoed the worry of others when he commented, "closing Letterman as a Military Hospital would leave me with nowhere to go. All of my medical records were at Letterman, all the doctors who knew me were at Letterman"; see Peter Tira, "Area's Military Vets Go to Battle over Presidio Hospital," *San Francisco Independent,* December 14, 1993: A1.

5. Dawn Garcia, "Presidio Park Advocates Look to Private Money," *San Francisco Chronicle,* January 16, 1989: A1.

6. Brian O'Neill, GGNRA superintendent, National Park Service, interview, San Francisco, May 28, 1996.

7. Alcatraz had a sinister reputation as a prison for some of America's most notorious criminals, including Al Capone, George "Machine Gun" Kelly, Alvin Karpis, Doc Barker, and Robert "Birdman" Stroud. Throughout the 1950s and 1960s, the cost of operating and maintaining Alcatraz had risen higher and higher. Attorney General Robert Kennedy decided to close the prison in March 1963. Alcatraz was then transferred to the federal government's General Services Administration (GSA). The GSA offered the island to various government agencies, but no one was interested. By 1968, without an interested federal agency, the GSA declared Alcatraz "surplus government property" and announced it would be put on the auction block. Meanwhile, local politicians and citizens argued for a new future for Alcatraz; see James Delgado, *Alcatraz: Island of Change* (San Francisco: Golden Gate National Park Association, 1991: 39). San Francisco mayor Joseph Alioto lobbied the GSA for the right to acquire the island, and he requested the community submit proposals. The city Planning Department received a flood of proposals from would-be developers who coveted the site for hotels, colleges, housing, a gambling casino, and even a "Western Statue of Liberty." Alioto chose to support a proposal submitted by Texas millionaire Lamar Hunt, which called for demolishing most structures and turning the island into a commercial site. Outraged citizens fought the proposal and requested the Department of the Interior to consider making Alcatraz part of the national park system. Once again, Philip Burton rode to the rescue. Both he and the National Park Service proposed that the GGNRA legislation include Alcatraz Island's 22 acres. This suggestion eventually won the support of city officials. The planning commissioner along with the mayor and other city officials lobbied to ensure that Alcatraz retained its open space. For an account, see Allan Jacobs, "1968: Getting Going, Staffing Up, Responding to Issues," in *Introduction to Planning History in the United States,* ed. Donald Krueckeberg (New Brunswick, N.J.: Center for Urban Policy Research, 1983: 235–257).

8. The Presidio's location and its long history as part of an urban open space have given it a special relationship with many San Franciscans, similar to New York-

ers' feelings about Central Park or Londoners' feelings about Regents Park. As a comparative illustration of the value of this urban open space, imagine what would happen if several of the 843 acres of Central Park were opened to residential or commercial development. Already, buildings that border the park take advantage of their location and view and charge much higher rates. One recent development project on Fifth Avenue, a renovated residential hotel now a condominium complex, advertised its penthouse (with a panoramic view of the Great Lawn and Reservoir) for $5.6 million. The April 14, 1996, issue of the *New York Times Magazine* contained ads for studios and one-bedroom suites at the Trump International on Park Avenue complete with "sweeping windows emphasizing dramatic views over Central Park, the River and the Skyline," priced from $250,000 to $1,295,000 (April 14, 1996: 171). If the value of these buildings is enhanced by "views of Central Park," what might the value of a single-family house situated *in* the center of Central Park be on the real estate market? Or, to extend this illustration, if by some geologic miracle, Central Park were elevated and oriented so that most of the park offered unobstructed views of the Brooklyn Bridge, the Statue of Liberty, and downtown, for what astronomical price might these view lots be sold? Governors Island, a forested island off the southern tip of Manhattan, is a Coast Guard base recently closed. New Yorkers are just beginning to contemplate the "staggering real estate and residential opportunities—a hundred and seventy-three acres of prime harbor-front property with views of Wall Street, landmark houses, a nine-hole golf course, and a Burger King" (as reported in "The Talk of the Town," *The New Yorker*, April 22, 1995: 37–38). Although Governors Island is smaller than the Presidio, its conversion has elicited the same type of fevered comments as the Presidio's.

    9. Allen Temko, "Time Is Now to Plan a Visionary Future," *San Francisco Chronicle*, May 15, 1989: A1.

    10. Ibid.

    11. The Army had banned mushroom picking in 1987. Mark Norton, a Presidio Heights resident and member of the Mycological Society, argued, "The long tradition of mushroom collecting has been suppressed and that is a blow to several ethnic groups who cherish mushrooms" (Bruce Bellingham, "Last Presidio Public Hearing," *Marina Times*, January 1994).

    12. Former Soviet president Mikhail Gorbachev toured the Presidio in May 1992. Said Gorbachev: "I think this Army base is a beautiful symbol of our hope to locate the Gorbachev Foundation here. . . . the interest and goals of the Gorbachev Foundation symbolize the potential for the Presidio as an international place of significance" (Michael McCabe, "Gorbachev Declares Presidio Perfect Site for His Foundation," *San Francisco Chronicle*, May 9, 1992: A1). The Gorbachev Foundation, an organization dedicated to strengthening democracy, humanitarian causes, and peace, opened its U.S. headquarters at the Presidio in 1993.

    13. Liz Lufkin, "About-Face Ideas for the Presidio," *San Francisco Chronicle*, February 17, 1989: B3.

14. Diana Walsh, "The Presidio Inspires Me to Dream," *San Francisco Examiner*, February 12, 1989: B1.

15. Kevin Starr, "The Presidio," *San Francisco Examiner, Image,* November 20, 1988: 72–76.

16. *Presidio Update* newsletter, 1992: 2. In his challenge for a grand vision, Mott referred to Daniel Burnham's statement: "Make no little plans; they have no magic to stir men's blood, and probably themselves will not be realized. Make big plans; aim high in hope and work, remembering that a noble logical diagram once recorded will never die"; quoted in Thomas Hines, *Daniel Burnham* (New York: Oxford University Press, 1974).

17. The Golden Gate National Park Association is one of sixty-five "cooperating" associations authorized by Congress to support research, interpretation, and conservation programs of the Park Service. Established in 1981, GGNPA operates book-sale outlets, runs tours of Alcatraz and Muir Woods, and manages and finances restoration projects and funds. In the early years, GGNPA employed three people and had an annual budget of approximately $300,000; by 1995, GGNPA had 120 employees and an annual budget of approximately $8 million (Greg Moore, GGNPA executive director, interview, San Francisco, May 23, 1996). GGNPA assisted the Park Service with key elements of the Presidio's planning process, including publishing communications material and funding essential studies.

18. During the conceptualization stage, GGNPA called the organization the Presidio Task Force. This was meant to be a "working title" only. GGNPA Executive Director Greg Moore recalled that they wanted an elegant-sounding name for the organization without sounding as if it were claiming some privilege of direct oversight. They couldn't use "commission" because there was the GGNRA Advisory Commission. With the assistance of a good thesaurus, the name Presidio Council was chosen.

19. James Harvey, local San Franciscan and CEO of TransAmerica Corporation, had served as vice chair of the National Park Foundation. His participation was important, for he understood how the Park Service operated and had a good understanding of Washington politics as well. Sadly, he died in 1997.

20. Toby Rosenblatt, chair, GGNPA, interview, San Francisco, May 28, 1996.

21. As stated in the Presidio Council's "Mission and Implementation Plan," June 1992 (document courtesy of William K. Reilly).

22. Among the Presidio pro bono partners were Arthur Andersen & Co. (an accounting firm that helped design a database system to compare future operating costs); Anshen+Allen (an architectural firm that analyzed the rehabilitation needs at several Presidio sites); Edelman Worldwide; Graphic Guides, Inc. / David Sibbert (contributed with design of Vision Workshops and the Presidio Forum); McKinsey & Company (advised on economic feasibility); Pacific Gas and Electric; TransAmerica Corporation and Urban Ecology (provided expertise on innovative technologies for Presidio maintenance and infrastructure repair). As reported by the National

Park Service in the *Draft General Management Plan Amendment: Environmental Impact Statement, Presidio of San Francisco* (San Francisco: GGNRA, October 1993).

23. *San Francisco Bay Guardian*, editorial, "The Presidio in Peril," January 12, 1994.

24. Among the members of the Advisory Commission were Amy Meyer and Edgar Wayburn, the original founders of the People for a GGNRA (PFGGNRA), the organization that had been instrumental in conceptualizing the GGNRA legislation in 1972.

25. The Master Plan of the City and County of San Francisco contains general land-use policies and objectives for San Francisco, including policies about recreation and open space that specifically mention the Presidio. In fact, the city had zoned the Presidio as "open space," a fact that would prove important in the battle for the Presidio in Congress. The GGNRA was obliged to regularly advise and seek comments from the city planning department on all planning matters related to the future use of the Presidio that could result in significant impacts on the physical, social, and economic environment of the city. For this reason, the appointment of a city planner to the Presidio planning team provided an important communication access.

26. O'Neill, interview.

27. Ibid. The Marina Middle School is located a few blocks away from the East Gate entrance to the Presidio.

28. Don Neubacher, superintendent, Point Reyes National Seashore (and former Presidio planning team captain), interview, Point Reyes, Calif., May 26, 1996; O'Neill, interview.

29. Richard Paddock, "View from Presidio: Profits," *Los Angeles Times*, August 12, 1992: A1.

30. O'Neill, interview; Moore, interview.

31. O'Neill, interview.

32. Ibid.

33. Dale Champion and Allen Temko, "Interior Secretary Sizes Up Presidio Park," *San Francisco Chronicle*, May 18, 1989: A3.

34. This general management plan did not address the management or administration of the Presidio, because it was still Army property.

35. National Park Service, *Creating a Park for the 21st Century: From Military Post to National Park. Draft General Management Plan Amendment, Presidio of San Francisco*, NPS D-148 (San Francisco: GGNRA, October 1993).

36. Dianne Feinstein, "Statement by Senator Dianne Feinstein, Hearing on Presidio Legislation (S. 1549 and S. 1639), Subcommittee on Public Lands, National Parks, and Forests," May 12, 1994: 6 (document obtained at the Senate hearing); U.S. Senate, *Management of the Presidio in San Francisco by the National Park Service, Hearing Before the Subcommittee on Public Lands, National Parks, and Forests of the Committee on Energy and Natural Resources, May 12, 1994* (Washington, D.C.: GPO, 1994: 4–10).

37. U.S. Senate, ibid.

38. This was not a unique determination. Since 1982, federal legislation had allowed private business to exist within national parks through the Historical Property Leasing Program. Under this program, commercial ventures lease historic buildings, thereby assuming responsibility for both maintenance and the buildings' historic integrity.

39. National Park Service, *Presidio Visions Kit* (San Francisco: GGNPA and GGNRA, Winter 1991).

40. Four workshops were held in San Francisco and one each in Marin, Alameda, Contra Costa, and San Mateo counties.

41. National Park Service, *Creating a Park for the 21st Century*, 122.

42. Neubacher, interview; O'Neill, interview.

43. In August 1992, the Park Service published "Summary of Responses to the Presidio Call for Interest," listing all the organizations that had indicated an interest in locating at the Presidio and submitted a response. The summary was sent to all respondents and was available to the public. It encouraged prospective tenants to contact organizations with similar interests and program ideas to facilitate more comprehensive proposals. Numerous organizations were interested in the Presidio. The majority of proposals came from nonprofit organizations. Among those that responded with suggestions highlighting the environmental stewardship and sustainability theme were the United Nations Environment Programme, the U.S. Environmental Protection Agency, the Gorbachev Foundation, the Commission on National and Community Service, the Youth Conservation Corps, and the University of California.

44. National Park Service, *Creating a Park for the 21st Century*, 123.

45. The plan was called a draft master plan, because there would be a public review period and perhaps revisions, followed by formal approval of the plan by the National Park Service.

46. The following list is a sample of the many types of surveys done for the Park Service as documented in National Park Service, *Draft General Management Plan Amendment: Environmental Impact Statement, Presidio of San Francisco*, NPS D-149 (San Francisco: GGNRA, October 1993): 1990, *Rare Plant Survey: Presidio of San Francisco*, by Jones & Stokes Associates, Inc.; 1992, *Historic Resource Study: El Presidio de San Francisco, a History Under Spain and Mexico, 1776–1846*, by John Phillip Langelier and Daniel B. Rosen; 1991, *Preliminary Engineering Study of the Electric Facilities for the Presidio of San Francisco*, by Jim Crane for PG&E; 1991, *Presidio Asbestos Abatement Survey*, by Ace Pacific Co.; 1992, *Water Supply Evaluation, Presidio of San Francisco*, by Nolte and Associates.

47. National Park Service, *Creating a Park for the 21st Century*, 20.

48. Ibid., 58.

49. Ibid., 56–59.

50. Several participants in the Letterman Design Group were also members of the Presidio Council.

51. I observed this discussion at the design group meeting I attended.

52. Neubacher, interview.

53. This information comes from several archival sources, including memos in William K. Reilly's personal archives, as well as personal observations.

54. National Park Service, *Creating a Park for the 21st Century*, 86.

55. Ibid., 82–85.

56. Ibid., 98.

57. Ibid., 106.

58. National Park Service, *Draft General Management Plan Amendment*, vii.

59. *National Parks and Conservation*, editorial, "The Presidio Update," January–February 1994: 10.

60. For a thorough analysis of this, see Richard DeLeon, *Left Coast City: Progressive Politics in San Francisco, 1975–1991* (Lawrence, Kans.: University of Kansas Press, 1992).

61. Gerald Adams, "Presidio Plan Wins Support at Unveiling," *San Francisco Chronicle*, October 20, 1993: A19–21.

62. Ibid., A21.

63. Martin Espinoza, "The Presidio Power Grab," *San Francisco Bay Guardian*, January 12, 1994: A1.

64. Moore, interview.

65. Some local citizens objected to the practice of animal research and previously had crusaded against the university's long-standing practice of using animals in biomedical research. They worried that the university might conduct animal research at the Presidio under the guise of conducting research to benefit human health.

Notes to Chapter 7

1. Harold Gilliam, "Sacred Ground," *San Francisco Examiner, This World,* June 19, 1994: 5–10.

2. From 1990 to 1992, the National Park Service was allocated only $3 million per year for the Presidio. Many in the community felt this amount was not sufficient because the Park Service was not only taking over many of the properties and duties (such as police and fire services) but also paying for the planning process. In response, President Bush approved an increase in the Park Service Presidio budget to $14.8 million for 1993. This was designed to help phase in Park Service operations at the post. It was generally assumed that in 1994, the year in which the Presidio was officially transferred to the Park Service, the budget would be $25 million, and that it would remain in the $25 million to $28 million range until the Presidio became

financially self-sufficient (with tenants in all the buildings, including the pivotal anchor tenant at the Letterman Complex) in 2010.

3. In all fairness, it should be noted that despite the shadow of defense downsizing, the Army had initiated a $70 million expansion program at the Presidio in 1988–1989. After the Presidio closure announcement, the Army finished construction of a new 96,000-square-foot commissary, new medical barracks at Letterman, and a new bowling alley. The Army probably could have abandoned these partially finished projects, but, to its credit, it did not.

4. Ingfei Chen, "Army Says It Goofed—OKs Money for Presidio," *San Francisco Chronicle*, February 12, 1992: A15.

5. Ingfei Chen, "Army Promises to Take Care of Presidio," *San Francisco Chronicle*, November 14, 1992: A14.

6. These studies included a Preliminary Assessment, the Remedial Investigation/Risk Assessment, the Feasibility Study, and the Remedial Action Plan (RAP). These studies were done by the Army and reviewed by the following regulatory agencies: U.S. EPA, California Department of Toxic Substances Control, and the Regional Water Quality Control Board. The latter two agencies were responsible for monitoring Army cleanup as part of the California state Superfund process.

7. A *San Francisco Chronicle* article recounted the story of a local ex-military base, not far from the Presidio, on Mount Tamalpais (a property within the GGNRA). In the 1970s, the Air Force had abandoned its small Air Force station as part of a reorganization process. On this 100-acre site sat forty structures, including barracks, a swimming pool, cracked basketball courts, and a row of houses (since inhabited by lizards and snakes). The buildings were filled with toxic asbestos, and jet fuel and cleaning solvents had contaminated the ground around many of the warehouses. The Park Service inherited the Air Force station as part of the GGNRA but was not given money to remove the asbestos, nor did the Department of Defense return to clean up the environmental contamination. Brian O'Neill, GGNRA superintendent, lamented, "the Air Force just threw the keys over the fence and said, 'It's your baby.'" The article concluded that this was an example of what might happen when military bases were closed without sufficient funds for cleaning up environmental contamination (the implied message: we must remain vigilant watchdogs about environmental cleanup at the Presidio because the Army might try to sneak out without completing the job).

8. For many years the EPA had the authority to penalize the private sector for mishandling hazardous waste, but it had only been granted the authority to oversee hazardous waste practices at federal facilities in 1992. The inspection of the Presidio was part of a broader campaign by the EPA to monitor practices at federal facilities.

9. Elliot Diringer, "EPA Fines Army for Mishandling Wastes at Presidio, *San Francisco Chronicle*, May 10, 1994: A2.

10. Nancy Pelosi, *Presidio Report from Congresswoman Nancy Pelosi, U.S. House of Representatives,* November 1993.

11. Miller and Vento became involved because Miller was chair of the House Natural Resources Committee and Vento was chair of the House National Parks Subcommittee.

12. *San Francisco Bay Guardian,* editorial, December 1993.

13. Among the select private members was the baseball great Joe DiMaggio.

14. Phillip Matier and Andrew Ross, "Army Wants to Stay the (Golf) Course," *San Francisco Chronicle,* November 22, 1993: A15.

15. *San Francisco Chronicle,* editorial, "The Army Role as a Tenant at the Presidio," *San Francisco Chronicle,* January 6, 1994.

16. Charles McCoy, "Astonishing Views, and Many Opinions; Must Be the Presidio," *Wall Street Journal,* April 19, 1994: A1.

17. GGNPA, *Presidio Update,* "An Update of the National Park Service Presidio Planning Process" (San Francisco: National Park Service/GGNRA, Summer 1994: 3).

18. Carl Nolte, "Army Trying to Keep Presidio Golf Course," *San Francisco Chronicle,* January 6, 1994: A1.

19. National Park Service, *Creating a Park for the 21st Century: From Military Post to National Park. Draft General Management Plan Amendment, Presidio of San Francisco,* NPS D-148 (San Francisco: GGNRA, October 1993: 110).

20. Robert Chandler, the National Park Service's Presidio general manager until 1997, explained: "We don't have the legal authority to do what a public corporation could do—to go to commercial markets and borrow money to upgrade facilities. We don't have the authority to retain revenues from leases. We're also constrained by government procurement practices that tend to be much more long-term and costly than what the private sector can do"; quoted in Harold Gilliam, "Private Interests in a Public Treasure," *San Francisco Examiner, This World,* May 15, 1994: 7.

21. There were several reasons why this form of management became an exciting alternative. Cutbacks at the federal level, industrial restructuring, globalization, and the promotion of President Reagan's New Federalism—which explicitly encouraged the enlargement of the public sector's role in helping cities adapt to this change—inspired a torrent of newspaper stories, magazine think pieces, and journal articles touting public-private partnerships as the way to save cities from declining budgets. The most visible downtown redevelopment projects—sponsored by partnerships in Philadelphia, Baltimore, Pittsburgh, Milwaukee, and New York—received extensive publicity; see R. S. Fosler and R. A. Berger, eds., *Public-Private Partnerships in American Cities* (Lexington, Mass.: Lexington Books, 1982). In many cases, partnerships were established to combat central-city decline due to currents such as suburbanization, the shift from a manufacturing to a service-based economy, and migration. Although the concept of public-private partnerships is not new (the civic

improvement associations of the City Beautiful era in the early 1900s and the 1940s Greater Pittsburgh Allegheny Conference for Community Development are earlier examples), public-private partnerships were "rediscovered" in the 1980s and embraced as a vehicle by which private capital could be attracted to an otherwise economically distressed city or community; see Max Stephenson, Jr., "Whither the Public-Private Partnerships: A Critical Overview," *Urban Affairs Quarterly* 27(1), 1991: 109–127. In 1982, the President's Committee for Economic Development released a report that called for the creation of urban partnerships. Since this report, there have been several studies detailing public-private partnerships in urban "redevelopment projects." Particular attention has been paid to the city of Pittsburgh and its Pittsburgh Renaissance partnership, which implemented projects such as industrial smoke abatement, traffic and bridge reconstruction, and downtown renewal projects (including a complex of hotels and office buildings that replaced warehouses, port facilities, and factories); see Fosler and Berger, *Public-Private Partnerships,* and the August 1997 issue of the *Journal of the American Planning Association,* which is devoted to a symposium on public-private partnerships in Pittsburgh. Although a few cities have established public-private partnerships to manage city parks (that is, they contract out lawn maintenance, tree trimming, and so on), in no national park has this concept been used as a solution to the complex challenges of park administration.

22. Perry Davis, ed., *Public-Private Partnerships: Improving Urban Life. Proceedings of the Academy of Political Science* 36(2) (New York: Academy of Political Science, 1986).

23. McKinsey & Co. donated their own resources for this study. Informal estimates place the value of this donation between $750,000 and $1 million. The study reinforced the credibility of the partnership concept because it indicated to many audiences that thorough, nonpartisan research had been completed.

24. Toby Rosenblatt, "Statement of Toby Rosenblatt, Vice Chair, Presidio Council, Chair, Golden Gate National Park Association," to the House Subcommittee on National Parks, Forests, and Public Lands, October 25, 1993: 7 (document obtained at the hearing). Originally, the Park Service referred to the Presidio public-private partnership as a public benefit corporation and gave it the name Presidio Corporation. I use the term public-private *partnership,* which better describes the intention and management structure.

25. The McKinsey & Co. report highlighted the nearby Fort Mason Foundation, which was smaller in scope but had the same public-spirited goals, and the Pennsylvania Avenue Development Corporation (in Washington, D.C.), which was the right size but had different goals. The report recommended that the Presidio public-private partnership be similar to the Fort Mason Foundation in its goals and similar in structure to the Pennsylvania Avenue Development Corporation. The Fort Mason Center, formerly military property, is a 13-acre waterfront complex that is now a conference center and a cultural cornucopia of theaters, museums, the acclaimed Greens restaurant, galleries, and office space for more than fifty nonprofit

groups (these groups include Friends of the River, Friends of the San Francisco Public Library, California Tomorrow, the Ploughshares Fund, and the National Poetry Association). Fort Mason is also part of the Golden Gate National Recreation Area, and thus many thought it could be an ideal case study for what should be done at the Presidio. The Fort Mason Foundation manages hundreds of tenants and activities and administers a $1.5 million yearly operating budget. The Fort Mason Foundation, however, is too small, not only geographically but also in budget, to provide an exact blueprint for the Presidio. The larger Pennsylvania Avenue Development Corporation used government money and private-sector expertise to generate more than $1 billion for redevelopment projects. Unlike many other public-private partnerships, the Pennsylvania Avenue Development Corporation could finance capital improvements through debt. This ability was considered crucial to the success of the Presidio, given the hundreds of buildings on post that needed rehabilitation and the sizable operating budget forecasted. McKinsey & Co. proposed using the structural model of the Pennsylvania Avenue Development Corporation with the Presidio Trust in order to secure this important allowance.

26. William K. Reilly, senior adviser to the Presidio Council, telephone interview, December 12, 1995.

27. Jerry Keyser, "Statement of Jerry Keyser, Keyser Marston Associates," to the House Subcommittee on National Parks, Forests, and Public Lands, October 26, 1993 (document obtained at the hearing).

28. Harvey had served as vice chair of the National Park Foundation. The chair of this organization is the secretary of the interior, who usually asks the vice chair to be the "active" chair. Harvey had valuable experience with how the Park Service operates, combined with a keen understanding of trends in federal government.

29. Toby Rosenblatt, chair, Golden Gate National Park Association Board of Directors, interview, San Francisco, May 28, 1996.

30. National Park Service, *Creating a Park for the 21st Century*, 111.

31. Ibid., 110.

32. Ibid.

33. Greg Moore, executive director, Golden Gate National Park Association, interview, San Francisco, May 23, 1996; William K. Reilly, senior adviser to the Presidio Council, telephone interview, April 5, 1996; Rosenblatt, interview.

34. Pelosi also proposed a second initiative, which would allow the Park Service to negotiate a tenant for the Letterman Complex. This was an important piece of legislation because it was considered an interim solution to the problem of leasing this area of the Presidio. The legislation gave the Park Service temporary authority to enter lease agreements until the Presidio Corporation was established and functioning. Normally, the National Park Service can negotiate leases only with concessionaires; any other type of lease agreement requires congressional approval. Pelosi's bill gave the Park Service the power to enter into a lease with the anchor tenant of the proposed Letterman Complex. Many experts considered the anchor

tenant for the Letterman Complex a key element of the Presidio park plan because of the revenue it would provide. The revenue generated from Letterman would be reinvested in the maintenance of other Presidio buildings and infrastructure. Robert Chandler, Presidio Park Service general manager, explained the need for the legislation: "We need the authority for Letterman because it's the most important part of the Presidio in terms of its revenue-generating capability, and it provides a way to get started with the leasing process. Without a lease by the time the Park Service takes over in October, the taxpayers would get a big bill for maintenance of the medical facilities there. Assuming that a corporation is established, the responsibility will be turned over to the corporation" (quoted in Gilliam, "Private Interests in a Public Treasure," 7). The legislation was attached as a rider to a bill dealing with parks in the state of Louisiana and was signed by President Bill Clinton in December 1993 (Public Law 103-175). As expected, by 1994 the Park Service had received sixteen submissions in response to the Request for Qualification to Lease Buildings in the Letterman Complex. After evaluation of these submissions, two were highlighted. The Tides Foundation proposed to convert 73,000 square feet of the Letterman General Hospital into offices, classrooms, and public exhibit space to establish the "Thoreau Center for Sustainability." And the University of California–San Francisco proposed to lease the entire 1.2 million-square-foot Letterman Complex to establish the "Presidio Center for Health Science Research and Education" (U.S. Department of the Interior, National Park Service, news release, "National Park Service Selects Tides Foundation and UCSF for Letterman Talks," June 21, 1994). The university's proposal was seen as pivotal and highly desirable.

35. Nancy Pelosi, "Pelosi Introduces Long-Term Presidio Bill," *News from Congresswoman Nancy Pelosi,* November 3, 1993.

36. Rosenblatt, "Statement."

37. According to Rosenblatt and Reilly, the Presidio Council endorsed the public-private partnership as a solution to Presidio management for four reasons. First, the partnership would have the flexibility to hire specialists in real estate management, finance, and leasing outside the federal service. It also would have the ability to finance capital improvements through both private and public borrowing. This ability was seen as pivotal to the operation of the Presidio since there were so many buildings in need of rehabilitation. The Presidio Corporation, unlike a public agency (such as the Park Service), would be allowed to take on debt to finance capital improvements. Second, the partnership would reduce decision turnaround time, creating more efficient management and saving money by leasing buildings more quickly than a government bureaucracy would be able to do. And, because the corporation would manage all built-up areas of the Presidio, it might save additional money through economies of scale. Third, creation of a partnership for the Presidio would benefit from strong relationships with the philanthropic community. Fourth, the partnership would be accountable to the public. All park programs would be required to be consistent with the general management plan for the Presidio and the program use concepts.

38. U.S. Senate, *Management of the Presidio in San Francisco by the National Park Service, Hearing Before the Subcommittee on Public Lands, National Parks, and Forests of the Committee on Energy and Natural Resources, October 25 and 26, 1993* (Washington, D.C.: GPO, 1993: 40).

39. Ibid., 41–42.

40. Barbara Boxer, "Statement of Senator Barbara Boxer, Senate Energy and Natural Resources Subcommittee on Public Lands, National Parks, and Forests (S. 1639)," *News from U.S. Senator Barbara Boxer,* May 12, 1994: 3.

41. U.S. Department of the Interior, Memo: Draft Comments on H.R. 3433, March 14, 1994 (document courtesy of William K. Reilly).

42. Some have referred to the political stance of the *Bay Guardian* as "left of the tabloids."

43. Joel Ventresca, "Keep the Presidio Ours," *San Francisco Independent,* January 18, 1994.

44. Martin Espinoza, "The Presidio Power Grab," *San Francisco Bay Guardian,* January 12, 1994: A1. As Espinoza notes, in part, this worry arose because the legislation would declare the Presidio Corporation to be "an essential public and governmental function, thereby exempt from all taxes and special assessments from the federal government, the State of California and even the City and County of San Francisco (H.R. 3433, section h.12.j.)."

45. Ibid.

46. Although this concern may have been valid, the fact is that within most national parks there is some private business. The concessionaires that run most of the public service facilities, campgrounds, restaurants, and rental services are private businesses that contract with the Park Service. Thus the presence of private business at the Presidio would not set a precedent at the national parks.

47. Moore, interview.

48. *San Francisco Examiner,* editorial, "Pelosi to Amend Plans for the Presidio," April 19, 1994: A6.

49. Ventresca's worry that the Presidio would become a commercialized, privatized business park was perhaps based on the assumption that public-private partnerships meant privatization. However, the Presidio Trust would not actually make a profit because the revenues generated would be returned to rehabilitate and maintain the structures on the post. By 1994, the majority of proposals for tenancy at the Presidio were from nonprofit (and noncommercial) organizations and included submissions from the California Academy of Science, the Exploratorium, the National Indian Justice Center, the Gorbachev Foundation, San Francisco State University, Stanford University, the Presidio Pacific Center, the San Francisco Conservation Corps, and the Sierra Club. Ventresca's criticisms of Presidio development seemed exaggerated. Local writer Harold Gilliam noted that while some of Ventresca's concerns were appropriate, he (Gilliam) felt the likelihood of commercialization at the Presidio was small: "So far, no applications from Disney or Ringling Brothers, Barnum and Bailey" (Gilliam, "Private Interests in a Public Treasure"). Not long after

Gilliam's defense of the Presidio Trust, Ventresca and Espinoza wrote an article condemning Gilliam for having been co-opted by corporate powers. Ventresca noted that Gilliam had been paid $7,500 by the GGNRA to write about the Marin Headlands. This fact, they argued, made Gilliam's defense of Pelosi and the Presidio Trust legislation "tainted" (Martin Espinoza, "The Gilliam Connection," *San Francisco Bay Guardian,* June 29, 1994: 20). The tirade against Gilliam, a well-known and respected environmental writer, seemed designed to communicate to Congress that there was local controversy over the Presidio. In fact, the continued attacks only highlighted the fact that the *Bay Guardian* and some of its staff were on the margins of community sentiment about the Presidio. Unfortunately for the Presidio, the *Bay Guardian* was successful at creating mischief. Congressional opponents took advantage of *Bay Guardian*'s stance and announced that the city was divided over the Presidio.

50. Nancy Pelosi, "Attempt to Slash Presidio Funding Fails," *News from Congresswoman Nancy Pelosi,* July 15, 1993.

51. Ironically, there was no mention that the size of the federal debt was due in large part to the enormous defense buildup of the Reagan years, or that this spiraling federal debt had come into being on the watch of two Republican presidents.

52. John Duncan, letter to Congress, House of Representatives, July 14, 1993 (document courtesy of the Golden Gate National Parks Association); *Congressional Record for the House of Representatives,* "Debate on Duncan Amendment to H.R. 2520 FY 94 Interior Appropriations Bill," July 15, 1993.

53. The Presidio Pet Cemetery, one of the more unusual areas in the Presidio, had been cited in several travel guides as among San Francisco's "weirdest" places. It was an easy target for Duncan's creative energies.

54. Duncan, letter to Congress.

55. *Congressional Record.*

56. Craig Middleton, director of government relations, Golden Gate National Park Association, interview, San Francisco, May 28, 1996.

57. Richard Paddock, "Global Center Plan Unveiled for Presidio," *Los Angeles Times,* October 1, 1993: A1.

58. Marc Sandalow, "House GOP Assails Cost of Presidio Plan," *San Francisco Chronicle,* October 27, 1993: A1.

59. The next chapter explores the general debate over recreation areas in greater detail and explains that for many years politicians have accused recreation areas of having only local, not national significance.

60. Craig Middleton, Presidio Council, Memo to William K. Reilly, "D.C. Trip Report, October 29, 1993" (document courtesy of William K. Reilly).

61. Marc Sandalow, "Babbitt Wearily Confident About Future of Presidio," *San Francisco Chronicle,* February 1, 1994: A3.

62. Ibid.

63. Memo to William K. Reilly.

64. Bruce Bellingham, "Call to Mobilize Against Duncan Presidio Plan," *Marina Times*, April 1994.

65. Brian O'Neill, superintendent, Golden Gate National Recreation Area, National Park Service, interview, San Francisco, May 28, 1996.

66. Draft document H.R. 4078. Like Pelosi's bill, Duncan's bill would have created a public-private partnership for the Presidio. Duncan noted, however, that his bill would "front load the project with enough money to enable the public corporation to get started." See Carl Nolte, "Congressman Tries to Block Presidio Plan," *San Francisco Chronicle*, March 10, 1994: A1.

67. Ibid.

68. U.S. Senate, *Management of the Presidio* (1993).

69. Bellingham, "Call to Mobilize."

70. Ibid.

71. Tom Steinstra, "Presidio Park Plan Stirring Heated Debate," *San Francisco Examiner*, April 10, 1994: D12. Duncan relied on figures from a 1993 General Accounting Office report that audited the costs of the proposed plan and the three alternatives. The GAO report concluded it was difficult to estimate how much the proposed plan would cost and estimated that one-time costs for activities such as infrastructure repair and upgrade, building rehabilitation, and environmental cleanup (a Defense Department responsibility) could range from $650 million to $1.2 billion, depending on the plan ultimately selected by the Park Service. Also, the GAO report noted, Presidio costs could be lower because of the pending Defense Department plan to keep the Sixth Army headquarters at the Presidio, but because this was still tentative, the GAO did not include revised figures. Duncan chose to use the highest estimate in a range of possibilities. See U.S. General Accounting Office, *Report on Transfer of the Presidio from the Army to the National Park Service. Testimony Before the Subcommittee on National Parks, Forests, and Public Lands, Committee on Natural Resources, House of Representatives, October 26, 1993* (Washington, D.C.: GPO, 1993).

72. By early 1994, the Park Service and the Presidio Council had revised expense estimates as a result of the Defense Department decision to retain the Sixth Army at the Presidio (the department would incur maintenance costs for several of the administrative buildings at the Main Post as well as for the housing units) and as a result of an external analysis that concluded that the original plan had overestimated costs. Their report concluded that the Presidio's cost would be $592 million over fifteen years (not $1 billion, as Duncan argued).

73. Steinstra, "Presidio Park Plan Stirring Heated Debate," D12.

74. William K. Reilly, "Statement of William K. Reilly," to the House Subcommittee on National Parks, Forests, and Public Lands, October 26, 1993: 5 (document obtained at the hearing).

75. John Jacobs, "Saving the Presidio: Politics in Review," *The Sacramento Bee*, May 1, 1994.

76. Middleton, interview; Moore, interview; Rosenblatt, interview.

77. Dianne Feinstein, "Statement by Senator Dianne Feinstein, Hearing on Presidio Legislation (S. 1549 and S. 1639), Subcommittee on Public Lands, National Parks, and Forests," May 12, 1994 (document obtained at the Senate hearing).

78. U.S. Senate, *Management of the Presidio in San Francisco by the National Park Service, Hearing Before the Subcommittee on Public Lands, National Parks, and Forests of the Committee on Energy and Natural Resources, October 25 and 26, 1994* (Washington, D.C.: GPO, 1994: 4).

79. Ibid., 10.

80. Craig Middleton, director of government relations for GGNPA, explained that the argument fizzled because of the way the federal budget is compartmentalized. There are approximately thirteen subcommittees on appropriations, and each one of these has its own pot, determined by the overall budget. Each chair has to divvy up expenses from within the pot. The Presidio went from one pot (Defense Department) to another pot (Interior Subcommittee on Appropriations). The latter pot was substantially smaller than the Defense Department budget of billions of dollars: $25 million was a mere drop in the Defense Department's pot, an expenditure of $25 million out of the Interior Department's $1.5 billion budget was huge, especially when compared with other items in that budget. The Interior Subcommittee on Appropriations includes the Park Service, all the other public lands, and agencies such as the National Endowment for the Arts and the National Endowment for the Humanities. For the subcommittee to come up with $25 million for the Presidio would be very difficult, especially since it was working with a "level budget" (little or no increase from the previous year). So the argument that the federal government was, in fact, saving money by having the Presidio as a park was not persuasive because no one counted the "overall pot" (Middleton, interview).

81. Thaai Walker, "1,500 March in Support of Presidio Park," *San Francisco Chronicle,* April 25, 1994: A15.

82. Sabin Russell, "Wilson Makes Plea in S.F. for Presidio Park," *San Francisco Chronicle,* April 23, 1994: A17.

83. *San Francisco Examiner,* editorial, "Don't Abandon the Presidio," May 8, 1994: A14.

84. Kevin Shelley, "The Plan to Sell the Presidio," *San Francisco Examiner,* March 24, 1994: A21.

85. *San Francisco Independent,* editorial, "Pelosi's Presidio Solution," May 3, 1994.

86. U.S. Senate, *Management of the Presidio* (1994: 30).

87. Jim Specht, "Senators Worry Whether Presidio Will Drain Budget," *Marin Independent Journal,* May 13, 1994: B5.

88. A filibuster today is different from James Stewart's filibuster in *Mr. Smith Goes to Washington.* A senator rarely stands on the Senate floor and talks on and on. Generally, a filibuster is over before it even begins. A senator threatens to filibuster. To defeat a filibuster, a senator must gather 60 votes agreeing to end debate. Without those 60 votes, the threat of filibuster is as good as a filibuster, and the bill is

never brought to vote. This happened to the Presidio. Supporters did not have enough votes to overcome the fact that various opponents had put a hold on the bill. Presidio advocates spent much time and energy scurrying around trying to convince Senator X to take his or her hold off the bill; once one hold was off, another would appear from a different senator. Finally, Robert Dole, R.-Kans., leader of the Republican majority, put a hold on the bill. The time constraints of the last days of the session effectively ended the Presidio Trust legislation's chance to pass.

In addition, some observers speculate that when the bill reached the Senate, Senator Feinstein did not push the Presidio Trust legislation with enough urgency. As the last months of the 103rd Congress drew to a close, a flurry of legislation was pushed through the House and Senate. Feinstein devoted her energies to getting Congress to pass the highly contentious California Desert Protection Act, the largest wilderness bill ever in the continental United States (and a bill she had worked to pass for several years). She had to overcome a Senate filibuster against the act but was successful in getting the bill passed shortly before Congress adjourned. The California Desert Protection Act created two new national parks, Death Valley and Joshua Tree, and, as with the Presidio, there was considerable debate about whether these new parks would deprive other parts of the National Park Service of needed money. Some believe that in the afterglow of victory over her bill, Feinstein hesitated to push for yet another national park bill, believing that Congress would balk at approving another expensive park for California.

89. Carl Nolte, "Presidio Soldiers Furl Their Colors," *San Francisco Chronicle*, October 1, 1994: A1, A17.

90. A brief review of civics and the legislative process may be useful. Each Congress meets for a period of two years. Each two-year period is composed of two one-year sessions. Legislation that is introduced but not passed in the first session (as was the Presidio legislation) can roll-over to the second session and be reconsidered. If the legislation fails to pass in the second session before Congress adjourns, the bill is "killed." When the new Congress convenes, the legislation must be reintroduced (it is not automatically reconsidered). It is given a new number (H.R. or S.). It may have to work its way through subcommittees and committees again in order to be voted on (although it may not be necessary to repeat hearings and testimony). When the 103rd Congress adjourned at the end of 1994 (concluding its second session), the Presidio Trust legislation had not been passed. Senate filibustering had killed the bill. When the 104th Congress began, Pelosi had to reintroduce the legislation for the Presidio Trust. The legislation thus had to work its way through the subcommittees and committees before it could (once again) be voted on by the House and then the Senate.

91. Timothy Egan, "Peacetime Vision for Presidio in Jeopardy," *New York Times*, February 2, 1995: A1, A14.

92. Ibid., A14.

93. The Department of Defense independently decided that the Sixth Army

was no longer needed. Several observers believed that the transition battles, especially the golf course issue, were the driving factor behind the decision to deactivate the Sixth Army, but it is difficult to confirm this suspicion.

94. Robert Chandler, National Park Service, Presidio general manager, interview, San Francisco, May 31, 1996; Rosenblatt, interview.

95. Chandler, interview.

96. Rosenblatt, interview.

97. The final legislation stipulated that the president appoint six board members (one was required to be a veteran). Three members were to be from the Bay Area. The members had to have knowledge and expertise in one of the following: city planning, finance, real estate, or natural resource conservation. The seventh board member would be the secretary of the interior.

98. Rosenblatt, interview.

99. A "rider" bill is legislation that is attached to another bill, although it is not necessarily related in theme or substance to the bill to which it is attached. Attaching controversial riders to popular legislation is a common practice in Congress. For example, the federal budget bill almost always contains unrelated and unpopular riders in the hopes that Congress, which in the end must approve a budget bill, will not bicker about (or notice) the supplemental legislation.

100. Middleton, interview.

101. Ibid.

102. The Presidio Trust legislation was the first bill (labeled "Title 1") included in this legislation, which also authorized federal spending of $17.5 million to acquire Sterling Forest along the border of New York and New Jersey, creation of the nation's first protected tall-grass prairie in Kansas, and an increase in the number of cruise ships visiting Glacier Bay in Alaska. In all, the omnibus parks bill was some 190 pages long.

103. Most of the trust members had been actively involved in the Presidio Trust legislation through nonprofit, nongovernment organizations such as the Presidio Council and GGNPA. Their appointment confirms the important role these groups played in securing passage of the trust legislation.

Notes to Chapter 8

1. Golden Gate National Park Association, *Gateways: Quarterly Newsletter* (San Francisco: GGNPA, Winter 1996).

2. Ibid.

3. Ibid., 5.

4. Horace Albright, as told to Robert Cahn, *The Birth of the National Park Service: The Founding Years, 1913–33* (Salt Lake City: Howe Brothers, 1985); Alston Chase, *Playing God in Yellowstone: The Destruction of America's First National Park* (San Diego: Harcourt Brace Jovanovich, 1987); Lary Dilsaver, ed., *America's National Park System:*

The Critical Documents (Lanham, Md.: Rowman and Littlefield, 1994); Ronald Foresta, *America's National Parks and Their Keepers* (Washington, D.C.: Resources for the Future, 1985); George Hartzog, Jr., *Battling for the National Parks* (Mt. Kisco, N.Y.: Moyer Bell, 1988); Dwight F. Rettie, *Our National Park System: Caring for America's Greatest Natural and Historic Treasures* (Urbana: University of Illinois Press, 1995); Hal Rothman, *Preserving Different Pasts: The American National Monuments* (Urbana: University of Illinois Press, 1989); Alfred Runte, *National Parks: The American Experience*, 2nd ed. (Lincoln: University of Nebraska Press, 1987); J. Douglas Wellman, *Wildland Recreation Policy: An Introduction* (Malabar, Fla.: Krieger Publishing Co., 1987); Conrad Wirth, *Parks, Politics and the People* (Norman: University of Oklahoma Press, 1980).

5. Rettie, *Our National Park System*, 14.

6. Dilsaver, *America's National Park System*.

7. Hans Huth, *Nature and the American: Three Centuries of Changing Attitudes* (Berkeley: University of California Press, 1957); Roderick Nash, *Wilderness and the American Mind* (New Haven, Conn.: Yale University Press, 1982); Joseph Petulla, *American Environmental History: The Exploitation and Conservation of Natural Resources* (San Francisco: Boyd and Fraser, 1977).

8. Some argue that Yosemite was the first national park. In 1864, Congress created Yosemite by transferring the land to the state of California as a protected reserve; however, it was not designated a national park until 1891.

9. See, for example, the discussion of Nash, *Wilderness and the American Mind;* and Max Oelschlaeger, *The Idea of Wilderness: From Prehistory to the Age of Ecology* (New Haven, Conn.: Yale University Press, 1991).

10. In *Wilderness and the American Mind,* Roderick Nash describes early American attitudes toward the wilderness. He concludes that in the early years of settlement, Americans feared the wilderness. They associated the wilderness with the fear of the unknown, chaos, amorality. At the close of the nineteenth century, however, Americans possessed a growing confidence in their ability to conquer or control nature. Confidence replaced fear and was part of the reason intellectual elites could reconstruct their perceptions of the wilderness and, by extension, the city.

11. Ibid., 44.

12. Ibid., 45.

13. Runte, *National Parks*, 11.

14. Nash, *Wilderness and the American Mind*.

15. Runte, *National Parks;* Rettie, *Our National Park System*.

16. Thomas Cox, "From Hot Springs to Gateway: The Evolving Concept of Public Parks, 1832–1976," *Environmental Review* 5(1), 1980: 20.

17. Simon Scharma writes, "The wilderness, after all, does not locate itself, does not name itself. It was an act of Congress." See *Landscape and Memory* (New York: Knopf, 1995).

18. Dilsaver, *America's National Park System;* Chase, *Playing God in Yellowstone;* Cox, "From Hot Springs to Gateway."

19. Hubert Work, "Statement of National Park Policy—Secretary Work, March 11, 1925," in Dilsaver, *America's National Park System*, 63. Runte, in *National Parks*, argues that early concerns about national parks focused on ways to increase and ensure tourism in the parks. It was commonly believed, even by many park supporters, that tourism and increased public use offered the best defense of park resources against development projects such as mining, logging, and grazing. As the park system evolved, the ideal of use and access would result in efforts to attract visitors to the parks by expanding railroads, building roads for automobiles, and providing infrastructure such as hotels. In the early years of the parks, few were concerned about the impact of visitors on the parks. The promotion of public use was not a real concern among park advocates because most parks were located in isolated, hard-to-reach areas.

20. Dilsaver, *America's National Park System*.

21. Rothman, *Preserving Different Pasts*.

22. Ibid., xi.

23. Runte, *National Parks*, 55.

24. This division of administration might lead some to disagree with my assertion that these cultural and historic sites were thought of as "national parks" in the years before the National Park Service existed. I would counter that these culture/historic places shared with the nature areas a fundamental purpose: to protect and preserve a place as a part of the American heritage. For this reason, I argue that they are national parks, even though decades passed before they were brought under the administration of the National Park Service.

25. Runte, *National Parks*, 104.

26. Cox, "From Hot Springs to Gateway."

27. Dilsaver, *America's National Park System*, 65.

28. Rettie, *Our National Park System*, 5.

29. Dilsaver, *America's National Park System*, 193–195.

30. Laurance Rockefeller, "A Report to the President and to Congress by the Outdoor Recreation Resources Review Commission, Laurance S. Rockefeller, Chairman, January 1962," ibid., 224.

31. Rockefeller, ibid., 227.

32. Cox, "From Hot Springs to Gateway," 20.

33. Lary Dilsaver, personal communication, annual Association of American Geographers conference, Fort Worth, April 1997.

34. Recall the Parks to the People program discussed in Chapter 4.

35. Recall that the driving force behind the early national park movement of the nineteenth century was antiurbanism. Not surprisingly, antiurban sentiment continued to underlie the parks in the twentieth century.

36. The tension between notions of the sublime and the picturesque is nicely illustrated by this example. A picturesque place does not inspire the stronger emotions associated with the sublime. Rather, the picturesque can be places of decay,

rough and charming rather than distinguished. For example, old barns can be picturesque, but the Grand Canyon is sublime. This excerpt from the autobiography of Lemuel Garrison, a Park Service ranger, makes the difference clear:

> I ride on horseback to the very top of the trail through Donahue Pass on the eastern boundary of Yosemite National Park and stop abruptly. Suddenly, I see to the far edge of the world in the eastern distance, across an endless jumble of wild mountain tops. Standing as tall as I can before the Lord, I am humbled and bareheaded and silent. I have met Creation.

See Lemuel Garrison, *The Making of a Ranger: Forty Years with the National Parks* (Salt Lake City: Howe Brothers, 1983: 300–301).

37. The national lakeshores and seashores also generated debate, but we will concentrate on the urban recreation areas.

38. Rettie, *Our National Park System*, 6.

39. Ibid., 32.

40. The Conservation Foundation has been one of the Park Service's most consistent support groups. It is a resource-oriented thinktank that has a reputation for political neutrality.

41. Conservation Foundation, *National Parks for the Future: An Appraisal of the National Parks as They Begin Their Second Century in a Changing America* (Washington, D.C.: Conservation Foundation, 1972: 8).

42. Ibid., 15.

43. Ibid.

44. Rettie, *Our National Park System*, 73.

45. Ibid., 74.

46. Ibid., 73.

47. This hierarchy owes a great deal to the suggestions and input of Lary Dilsaver, who spoke with me at length at the Association of American Geographers conference in Fort Worth in April 1997. He pointed out that recreation areas have neither nature nor culture characteristics.

48. Roger Kennedy, "Preamble to the Presidio Plan," press release, U.S. Department of the Interior, October 8, 1993.

49. Rettie, *Our National Park System*.

50. The issue of entrance fees at parks is of little consequence here. They have been either nonexistent or very affordable. Rather the issue was that the working class had neither the time to travel great distances nor necessarily the means to purchase transportation to the parks.

51. Rettie, *Our National Park System*, 50.

52. Supposedly, many park managers had come to realize the problems in relying on classifications and had given these distinctions less emphasis. But the categories continued to be evident in the titles used for park areas and, more important, in determining the budget and allocation of money and staff (ibid., 42). Despite the

articulation of the wholeness of the system, in 1972 under National Park Service Director George Hartzog, the Park Service created a thematic plan. Ironically, the plan consisted of two major categories: historic/culture resources and natural-area resources. Within these two broad categories were dozens of themes to which a site could contribute (examples of significance with regard to "telegraph and telephone" [the Pony Express Terminal in California], or "examples of Eolian landforms"). In all fairness, this was an attempt to measure how well the parks represented the diverse facets of American history, but it is notable that the major division of nature/culture persisted as a difficult dilemma to reconcile, even among those who embraced the concept of a single system.

53. Park historians have noted that the National Park Service has rarely reacted negatively to any proposed park or recreation area that has strong congressional support. By custom, the National Park Service responds favorably to proposals from influential congressional representatives like Burton.

54. Dilsaver, *America's National Park System;* Rettie, *Our National Park System.*

55. Rettie, ibid.; James M. Ridenour, *The National Parks Compromised: Pork Barrel Politics and America's Treasures* (Merriville, Ind.: ICS Books, 1994).

56. Runte, *National Parks,* 79.

57. Rettie, *Our National Park System,* 1.

58. The recommendations of the Conservation Foundation in its 1972 report testify to the influence environmental organizations have had on shaping or refocusing national park ideals. As previously mentioned, the Conservation Foundation felt that the Park Service needed to return to the original mission of conservation (however inappropriate that may be, given the historic diversity of the parks).

59. Dilsaver, *America's National Park System,* 373.

60. Ibid., 406.

61. The 1978 addition of 40 million acres of Alaskan wilderness contributed greatly to this figure.

62. National Park Service, *National Parks for the 21st Century: The Vail Agenda* (Post Mills, Vt.: Chelsea Green Publishing/National Park Service, 1993: xiii), hereafter cited as *Vail Agenda.* See also Ridenour, *The National Parks Compromised.*

63. National Park Service, *Vail Agenda.*

64. To paraphrase Roger Kennedy, director of the Park Service: self-scrutiny and self-renewal are rare enough among individuals, let alone bureaucracies (ibid., xiii).

65. Ridenour, *The National Parks Compromised,* 71.

66. Rettie, *Our National Park System,* 14.

67. Ibid., 2.

68. Cox, "From Hot Springs to Gateway," 14.

69. Rettie, *Our National Park System,* 25.

70. Ibid., 27.

71. National Park Service, *Vail Agenda,* 14.

72. Ibid., 3.

73. Joseph Penfold, Stanley Cain, Richard Estep, Brock Evans, Roderick Nash, Douglas Schwartz, and Patricia Young, "Preservation of National Park Values: Report of a Task Force Assembled by the Conservation Foundation," in Conservation Foundation, *National Parks for the Future*, 31.

74. Rettie, *Our National Park System*, 1.

75. Ridenour, *The National Parks Compromised*, 17. Ridenour explained that since parks don't generally have predators, few are ever decommissioned. The more marginal parks created, the thinner the blood of the system becomes.

76. Foresta, *America's National Parks and Their Keepers*, 280–281.

77. Quoted in Rettie, *Our National Park System*, 76.

78. Ridenour, *The National Parks Compromised*, 107–108.

79. Ibid., 17.

80. Greg Moore, executive director, Golden Gate National Park Association, interview, San Francisco, May 23, 1996.

81. Rettie, *Our National Park System*, 77.

82. I do not have the name of the individual who made this remark during a meeting I attended in October 1994.

83. Brian O'Neill, superintendent, Golden Gate National Recreation Area, National Park Service, interview, San Francisco, May 28, 1996.

84. U.S. Senate, *Management of the Presidio of San Francisco by the National Park Service, Hearing Before the Subcommittee on Public Lands, National Parks, and Forests of the Committee on Energy and Natural Resources, October 25 and 26, 1994* (Washington, D.C.: GPO, 1994: 39).

85. Moore, interview.

Notes to Chapter 9

1. See David Nicholson-Lord, *The Greening of the Cities* (London: Routledge & Kegan Paul, 1987), for a definition.

2. Peter Hall, *Cities of Tomorrow* (New York: Basil Blackwell, 1988); Peter Hall, *Great Planning Disasters* (Berkeley: University of California Press, 1980).

3. National Park Service, *Creating a Park for the 21st Century: From Military Post to National Park. Draft General Management Plan Amendment, Presidio of San Francisco*, NPS D-148 (San Francisco: GGNRA, October 1993).

4. William H. Wilson, "Moles and Skylarks," in *Introduction to Planning History in the United States*, ed. Donald Krueckeberg (New Brunswick, N.J.: Center for Urban Policy Research, 1983: 101).

5. Wendell Bell and James Mau, eds., *The Sociology of the Future: Theory, Cases and Annotated Bibliography* (New York: Russell Sage Foundation, 1971: 18).

6. Fred Polak, *The Image of the Future: Enlightening the Past, Orienting the Present, Forecasting the Future*, vol. 1 (New York: Oceanna Publications, 1961: 49–50).

7. Wilson, "Moles and Skylarks," 113.

8. Robert Chandler, National Park Service, Presidio general manager, interview, San Francisco, May 31, 1996; Brian O'Neill, superintendent, Golden Gate National Recreation Area, National Park Service, interview, San Francisco, May 28, 1996; Toby Rosenblatt, chair, Golden Gate National Park Association Board of Directors, interview, San Francisco, May 28, 1996.

9. Dwight F. Rettie, *Our National Park System: Caring for America's Greatest Natural and Historic Treasures* (Urbana: University of Illinois Press, 1995: 118). Rettie notes, for example, the general management plan for Gateway NRA in New York and New Jersey took ten years to approve; the plan for Yosemite took over six years. By comparison the Presidio plan was completed rather quickly, though not quickly enough to take advantage of the political momentum for the plan in Congress.

10. Lary Dilsaver, review of Ridenour's *The National Parks Compromised* and Rettie's *Our National Park System*, *Annals of the Association of American Geographers* 86(2), 1996: 357.

11. This source requested anonymity.

12. Rosenblatt, interview.

13. Brian Huse, director, Pacific Regional Office, National Parks and Conservation Association, interview, Oakland, Calif., May 28, 1996.

14. Joel Ventresca, "Keep the Presidio Ours," *San Francisco Independent,* January 18, 1994.

15. Rettie, *Our National Park System,* 117.

16. William K. Reilly, telephone interview, December 12, 1995.

17. Greg Moore, executive director, Golden Gate National Park Association, interview, San Francisco, May 23, 1996.

18. Alfred Runte, *National Parks: The American Experience,* 2nd ed. (Lincoln: University of Nebraska Press, 1987: 137–145).

19. Craig Middleton, director of government relations, Golden Gate National Park Association, interview, San Francisco, May 28, 1996.

20. Huse, interview.

21. O'Neill, interview.

22. Chandler, interview.

23. Ibid.

24. See Neil Everden, *The Social Creation of Nature* (Baltimore: Johns Hopkins University Press, 1992); Carolyn Harrison and Jacquelin Burgess, "Social Constructions of Nature: A Case Study of Conflicts over the Development of Rainham Marshes," *Transactions of British Geographers* 19, 1994: 291–310; Cindi Katz and Andrew Kirby, "In the Nature of Things: The Environment and Everyday Life," *Transactions of British Geographers* 16, 1991: 259–271; Michael Soulé and Gary Lease, eds., *Reinventing Nature? Responses to Postmodern Deconstruction* (Washington, D.C.: Island Press, 1995); Alexander Wilson, *The Culture of Nature: North American Landscape from Disney to Exxon Valdez* (Cambridge, Mass.: Blackwell, 1992); Will Wright, *Wild Knowledge: Science, Language, and Social Life in a Fragile Environment* (Minneapolis: University of Minneapolis Press, 1992).

25. Margaret FitzSimmons claims that geographers have displayed a "peculiar silence" on the question of nature and the environment. She contends that until recently nature had been left to the physical geographers; human geographers looked to either science (favoring economics, sociology, or a positivist social science) or history (working within a form of historical reductionism). Human geographers explored the structure of spaces but shied away from questions of human-nature relationships. See Margaret FitzSimmons, "The Matter of Nature," *Antipode* 21(2), 1989: 106–120. Her argument is persuasive, but there are of course exceptions: Clarence Glacken, *Traces on the Rhodian Shore: Nature and Culture in Western Thought from Ancient Times to the End of the Eighteenth Century* (Berkeley: University of California Press, 1967); Donald W. Meinig, ed., *The Interpretation of Ordinary Landscapes* (New York: Oxford University Press, 1979); and Yi-Fu Tuan, *Topophilia: A Study of Environmental Perception, Attitude and Values* (Englewood Cliffs, N.J.: Prentice-Hall, 1974), are cultural and historical geographers who have addressed these issues. Although geographers have looked at perceptions, attitudes, and values about the environment, few have addressed the impact of the environment (*qua* environmental movement) as a social and *political* force. This subject has become a topic in which more geographers are doing research. It is perhaps this fine line to which FitzSimmons refers when she argues that human geographers have been silent on nature. In the past, the term *environmental geographer* has been appropriated by physical scientists. I believe the term needs to be inclusive of the many ways in which scholars explore the environment; for example, I consider myself an environmental geographer, although my training is not in earth sciences but in the social sciences.

26. Carolyn Merchant, *The Death of Nature: Women, Ecology and the Scientific Revolution* (London: Wildwood House, 1980); Max Oelschlaeger, *The Idea of Wilderness: From Prehistory to the Age of Ecology* (New Haven, Conn.: Yale University Press, 1991).

27. Everden, *The Social Creation of Nature;* Oelschlaeger, ibid. In addition, these dualisms do manifest themselves in concrete ways. Consider, for example, the division within most universities in the United States between natural sciences and humanities. The discipline of geography itself has long been fragmented into physical geography and human geography.

28. Everden, ibid.; Oelschlaeger, ibid.

29. John Rennie Short, *Imagined Country: Society, Culture and Environment* (London: Routledge, 1991).

30. Wright, *Wild Knowledge*, xii.

31. Everden, *The Social Creation of Nature*, 93.

32. I am, however, reluctant to extend the logic of this argument to its most extreme and assert that since nature is a cultural construction, it must be illusory. If it is illusory, then everything is thus artificial. This type of relativism would insist that Disney's wild jungle ride is just as legitimate as the Grand Canyon. Although I find the deconstructionist perspective helpful in acknowledging that we perceive and define nature in cultural terms, this is not the same as saying there is no nature. In

Reinventing Nature? Soulé and Lease argue persuasively against this deconstructionist nihilism. They argue that—unlike Disneyland, plastic trees, or indoor downhill ski runs—many places possess a prescience and inspire a continuity with the surrounding world. I would posit that there is an actual context and structure of nature, and I would agree with Soulé and Lease and others who have argued that flora, fauna, and ecosystems have been determined by geographic location and geology more than by human activity or imagination. This would indicate that there is an "empirical" nature and a "constructed" nature. In this book, I focus more on conceptual nature and the cultural context in which humans understand and describe nature.

 33. See Glacken, *Traces on the Rhodian Shore;* Huth, *Nature and the American.*

 34. Roderick Nash, *Wilderness and the American Mind* (New Haven, Conn.: Yale University Press, 1982: xii).

 35. Alexis de Tocqueville, *Democracy in America* (New York: Harper and Row, 1966).

 36. Peter Schmitt, *Back to Nature: The Arcadian Myth in Urban America* (Baltimore: Johns Hopkins University Press, 1989: xx–xxi).

 37. Numerous examples exist, including those who assert that certain national parks (Yellowstone, Alaskan Wilderness, Yosemite) are pristine or unspoiled.

 38. Rettie, *Our National Park System.*

 39. Thomas Cox, "From Hot Springs to Gateway: The Evolving Concept of Public Parks, 1832–1976," *Environmental Review* 5(1), 1980: 20.

 40. Rettie, *Our National Park System.*

 41. Dilsaver, review. It could be argued that to reverse the designation of a national park is to devalue a previous generation's expression of what they considered significant. The decommissioning of a park is a dangerous action to take, because it places one generational value of nature and parks over another. Neither would I want to argue cultural relativism and say that all decisions are equal. However, I would argue that the creation of national parks is a spatial manifestation of place-significance. Reversing a decision about a park can mean the irrevocable loss of that place from the public domain, since many parks, once decommissioned, revert to private ownership. For any one group of people to decide that a place no longer deserves to be a national park may set in motion many irreversible changes (in both meaning and the physical landscape). The decommissioning of any national park, therefore, is not merely a change in the purpose and meaning of that place; it can mean the permanent loss of that space to the public commons. Most decisions to set aside or protect land from development are lauded today. For example, few people in New York regret the decision to create Central Park, and fewer still look at the preservation of Central Park and see it as a loss of potential profit from what certainly would have been commercial or residential development.

 42. See David M. Graber, "Resolute Biocentrism: The Dilemma of Wilderness in National Parks," in Soutlé and Lease, *Reinventing Nature?* 132.

 43. See, for example, William Cronon, *Changes in the Land: Indians, Colonists*

and the Ecology of New England (New York: Hill and Wang, 1983); and William M. Denevan, "The Pristine Myth: The Landscape of the Americas in 1492," *Annals of the Association of American Geographers* 82(3), 1992: 369–385. Also consider the debate about fire suppression and fire control policy in the national parks. For many years, the practice at many national parks was to prevent fires from starting or to put out fires as soon as possible. This, in effect, was a human-imposed control of a natural process. The 1988 Yellowstone fires, which burned hundreds of thousands of acres, sparked debate about the wisdom and practice of fire suppression in the parks. Although unintended, the policy actually contributed to more destruction of forest than would have occurred had natural fires been allowed to burn. Without periodic fires, large amounts of branches, tree limbs, and brush accumulate on the forest floor. When the Yellowstone fires began, there was a wealth of dry wood to burn as kindling. The impact of other park policies on the physical environment is well documented in Alston Chase, *Playing God in Yellowstone: The Destruction of America's First National Park* (San Diego: Harcourt Brace Jovanovich, 1987).

44. Denevan, "The Pristine Myth."

45. John Rennie Short, *The Humane City: Cities as If People Really Mattered* (Oxford: Basil Blackwell, 1989); M. White and L. White, *The Intellectual Versus the City* (Cambridge, Mass.: Harvard University Press and MIT Press, 1962).

46. Short, ibid.

47. Martin Melosi, "The Place of the City in Environmental History," *Environmental History Review* 17(1), 1993: 1–23.

48. Ibid., 4.

49. It should be acknowledged that cities with several national historic sites or national monuments have long worked closely with the Park Service. The Presidio does not represent the first attempt at reconciliation, only the most recent.

50. Nicholson-Lord, *The Greening of the Cities;* see also David Gordon, ed., *Green Cities: Ecologically Sound Approaches to Urban Space* (Montreal: Black Rose Books, 1990).

51. Nicholson-Lord, ibid.

52. Peter Berg, "A Green City Program with a Bioregional Perspective: Developing the San Francisco Green City Plan," in Gordon, *Green Cities,* 281–288.

53. Harold Gilliam, "View from the Presidio," *San Francisco Chronicle, This World,* June 16, 1991: 12.

54. City officials in Philadelphia, for example, were interested in the Presidio project as a possible template for how to plan for the conversion of the Philadelphia Naval Shipyard, located on the Delaware River in Philadelphia.

References

Note: The references are organized in three sections, each alphabetically arranged. The first section is general literature. The second section contains references to public documents—congressional hearings, National Park Service reports and plans, and so on. The third section lists the interviews referred to in the text.

Literature

Adams, Gerald. 1993, October 20. "Presidio Plan Wins Support at Unveiling." *San Francisco Chronicle*, A19–21.

Adams, Gerald. 1989, April 23. "Presidio Wish List Covers Gamut of Use." *San Francisco Examiner*, B1.

Agnew, John, John Mercer, and David Sopher, ed. 1984. *The City in Cultural Context*. Boston: Allen & Unwin.

Albright, Horace, as told to Robert Cahn. 1985. *The Birth of the National Park Service: The Founding Years, 1913–33*. Salt Lake City: Howe Brothers.

Albright, Horace. 1971. *Origins of National Park Service Administration of Historic Sites*. Philadelphia: Eastern National Park and Monument Association.

Ambrose, Stephen. 1976. *Rise to Globalism: American Foreign Policy, 1938–1976*. New York: Penguin.

Bell, Wendell, and James Mau, ed. 1971. *The Sociology of the Future: Theory, Cases and Annotated Bibliography*. New York: Russell Sage Foundation.

Bellingham, Bruce. 1994, April. "Call to Mobilize Against Duncan Presidio Plan." *Marina Times*.

Bellingham, Bruce. 1994, January. "Last Presidio Public Hearing." *Marina Times*.

Bensel, R. F. 1984. *Sectionalism and American Political Development, 1880–1980*. Madison: University of Wisconsin Press.

Berg, Peter. 1990. "A Green City Program with a Bioregional Perspective: Developing the San Francisco Green City Plan." In *Green Cities: Ecologically Sound Approaches to Urban Space,* edited by David Gordon, 281–288. Montreal: Black Rose Books.

Berger, John. 1941. *The Franciscan Missions of California.* Garden City, N.Y.: Doubleday.

Bolton, Herbert Eugene, ed. 1966 (1926). *The Historical Memoirs of New California by Fray Francisco Palóu, O.F.M.* 4 vols. New York: Russell and Russell.

Bolton, Herbert Eugene, ed. 1933. *Font's Complete Diary: A Chronicle of the Founding of San Francisco.* Berkeley: University of California Press.

Bolton, Herbert Eugene, ed. 1930. *Anza's California Expedition.* 5 vols. Berkeley: University of California Press.

Brandon, Henry, ed. 1992. *In Search of a New World Order: The Future of U.S.-European Relations.* Washington, D.C.: Brookings Institution.

Broek, Jan. 1965. *Geography: Its Scope and Spirit.* Columbus, Ohio: Merrill.

Brown, Neil. 1993, July 10. "Base-Closing Process Thwarts Parochialism." *Congressional Quarterly,* 1842.

Brugmann, B., and G. Stetteland. 1971. *The Ultimate Highrise.* San Francisco: SF Bay Guardian Books.

Budiansky, Stephen. 1989, November 20. "And the Wall Came Tumbling Down." *U.S. News & World Report,* 9.

Burnham, Daniel J. 1905. *Report on a Plan for San Francisco.* San Francisco: Sunset Press.

Caen, Herb. 1994, February 2. "Troublemaker." *San Francisco Chronicle,* B1.

Caen, Herb. 1976. *One Man's San Francisco.* Sausalito, Calif.: Comstock Editions.

Capron, Elisha Smith. 1854. *History of California.* Boston: n.p.

Carroll, Jerry. 1991, January 16. "Saviors of the Land." *San Francisco Chronicle,* B3.

Castells, Manuel. 1983. *The City and the Grassroots: A Cross-Cultural Theory of Urban Social Movements.* Berkeley: University of California Press.

Champion, Dale. 1988, December 30. "If the Army Pulls Out, S.F. Gets a Huge Park." *San Francisco Chronicle,* A13.

Champion, Dale, and Allen Temko. 1989, May 18. "Interior Secretary Sizes Up Presidio Park." *San Francisco Chronicle,* A3.

Chase, Alston. 1987. *Playing God in Yellowstone: The Destruction of America's First National Park.* San Diego: Harcourt Brace Jovanovich.

Chen, Ingfei. 1992, November 14. "Army Promises to Take Care of Presidio." *San Francisco Chronicle,* A6.

Chen, Ingfei. 1992, February 12. "Army Says It Goofed—OKs Money for Presidio." *San Francisco Chronicle,* A15.

Cherny, Robert W. 1995. "City Commercial, City Beautiful, City Practical: The San Francisco Visions of William C. Raboth, John Phelan and Michael M. O'Shaughnessy." *California History* 73(4): 297–307.

Chomsky, Noam. 1994. *World Orders: Old and New.* New York: Columbia University Press.

Clary, Raymond. 1980. *The Making of Golden Gate Park: The Early Years, 1865–1906.* Sausalito, Calif.: California Living Books.

Cole, Tom. 1981. *A Short History of San Francisco.* San Francisco: Don't Call It Frisco Press.

Committee for Economic Development. 1982. *Public-Private Partnership: An Opportunity for Urban Communities.* New York: Research and Policy Committee.

Conservation Foundation. 1972. *National Parks for the Future: An Appraisal of the National Parks as They Begin Their Second Century in a Changing America.* Washington, D.C.: Conservation Foundation.

Cosgrove, Denis, and Stephen Daniels, eds. 1988. *The Iconography of Landscape: Essays on the Symbolic Representation, Design and Use of Past Environments.* Cambridge: Cambridge University Press.

Cox, Thomas. 1980. "From Hot Springs to Gateway: The Evolving Concept of Public Parks, 1832–1976." *Environmental Review* 5(1): 14–26.

Cronon, William. 1991. *Nature's Metropolis: Chicago and the Great West.* New York: Norton.

Cronon, William. 1983. *Changes in the Land: Indians, Colonists and the Ecology of New England.* New York: Hill and Wang.

Dana, Richard Henry. 1840. *Two Years Before the Mast.* Roslyn, N.Y.: Classics Club.

Davis, Perry, ed. 1986. *Public-Private Partnerships: Improving Urban Life.* Proceedings of the Academy of Political Science 36(2). New York: Academy of Political Science.

DeGrasse, Robert. 1983. *Military Expansion, Economic Decline: The Impact of Military Spending on U.S. Economic Performance.* Armonk, N.Y.: M. E. Sharpe, Council on Economic Priorities.

DeLeon, Richard. 1992. *Left Coast City: Progressive Politics in San Francisco, 1975–1991.* Lawrence: University Press of Kansas.

Delgado, James. 1991. *Alcatraz: Island of Change.* San Francisco: Golden Gate National Park Association.

DePledge, Derrick. 1994, June 30. "House Committee Approves New Plan for Presidio Trust." *San Francisco Examiner,* A5.

Denevan, William M. 1992. "The Pristine Myth: The Landscape of the Americas in 1492." *Annals of the Association of American Geographers* 82(3): 369–385.

Dilsaver, Lary. 1996. Review of Ridenour's *The National Parks Compromised* and Rettie's *Our National Park System. Annals of the Association of American Geographers* 86(2): 355–357.

Dilsaver, Lary, ed. 1994. *America's National Park System: The Critical Documents.* Lanham, Md.: Rowman and Littlefield.

Diringer, Elliot. 1994, May 10. "EPA Fines Army for Mishandling Wastes at Presidio." *San Francisco Chronicle,* A2.

Duncan, James, and Nancy Duncan. 1988. "(Re)reading the Landscape." *Environment and Planning D: Society and Space* 6: 117–126.

Egan, Timothy. 1995, February 2. "Peacetime Vision for the Presidio in Jeopardy." *New York Times*, A1, A14.

Espinoza, Martin. 1994, June 29. "The Gilliam Connection." *San Francisco Bay Guardian*, 20.

Espinoza, Martin. 1994, January 12. "The Presidio Power Grab." *San Francisco Bay Guardian*, A1.

Everden, Neil. 1992. *The Social Creation of Nature*. Baltimore: Johns Hopkins University Press.

Featherstone, Mike, ed. 1990. *Global Culture*. London: Sage.

Feldman, Jonathan. 1988, May. *An Introduction to Economic Conversion: Briefing Paper One*. Washington, D.C.: National Commission for Economic Conversion and Disarmament.

Felton, John, and Pat Towell. 1988, December 17. "Bush Inherits Political Opportunities, Risks." *Congressional Quarterly*, 3503.

FitzSimmons, Margaret. 1989. "The Matter of Nature." *Antipode* 21(2): 106–120.

Foresta, Ronald. 1985. *America's National Parks and Their Keepers*. Washington, D.C.: Resources for the Future.

Fosler, R. S., and R. A. Berger, eds. 1982. *Public-Private Partnerships in American Cities*. Lexington, Mass.: Lexington Books.

Gaddis, John Lewis. 1992. *The United States and the End of the Cold War: Implications, Reconsiderations, Provocations*. New York: Oxford University Press.

Garcia, Dawn. 1989, January 16. "Presidio Park Advocates Look to Private Money." *San Francisco Chronicle*, A1.

Garrison, Lemuel. 1983. *The Making of a Ranger: Forty Years with the National Parks*. Salt Lake City: Howe Brothers.

Getlin, Josh. 1988, October 13. "Congress Passes Plan to Cut Bases." *Los Angeles Times*, A1.

Gilliam, Harold. 1994, June 19. "Sacred Ground." *San Francisco Examiner, This World*, 5–10.

Gilliam, Harold. 1994, May 15. "Private Interests in a Public Treasure." *San Francisco Examiner, This World*.

Gilliam, Harold. 1991, June 16. "View from the Presidio." *San Francisco Chronicle, This World*, 12–14.

Gilliam, Harold. 1967. *The Natural World of San Francisco*. Garden City, N.Y.: Doubleday.

Gilliam, Harold. 1957. *San Francisco Bay*. Garden City, N.Y.: Doubleday.

Glacken, Clarence. 1967. *Traces on the Rhodian Shore: Nature and Culture in Western Thought from Ancient Times to the End of the Eighteenth Century*. Berkeley: University of California Press.

Golden Gate National Park Association. 1994. *Presidio Gateways: Views of a National Historic Landmark at San Francisco's Golden Gate*. San Francisco: Chronicle Books.

Gordon, David, ed. 1990. *Green Cities: Ecologically Sound Approaches to Urban Space.* Montreal: Black Rose Books.

Gordon, Suzanne, and Dave McFadden, eds. 1984. *Economic Conversion: Revitalizing America's Economy.* Cambridge, Mass.: Ballinger.

Graber, David M. 1995. "Resolute Biocentrism: The Dilemma of Wilderness in National Parks." In *Reinventing Nature? Responses to Postmodern Deconstruction,* edited by Michael Soulé and Gary Lease, 123–136. Washington, D.C.: Island Press.

Hall, Peter. 1988. *Cities of Tomorrow.* New York: Basil Blackwell.

Hall, Peter. 1980. *Great Planning Disasters.* Berkeley: University of California Press.

Hall, Peter, and Ann Markusen. 1992. "The Pentagon and the Gunbelt." In *The Pentagon and the Cities,* edited by Andrew Kirby, 53–76. Newbury Park, Calif.: Sage.

Halle, Louis J. 1991. *The Cold War as History.* New York: HarperCollins.

Hampton, Duane. 1971. *How the U.S. Cavalry Saved Our National Parks.* Bloomington: Indiana University Press.

Harlow, Neal. 1982. *California Conquered: War and Peace in the Pacific, 1846–1850.* Berkeley: University of California Press.

Harraway, Donna. 1991. *Simians, Cyborgs and Women: The Reinvention of Nature.* New York: Routledge.

Harrison, Carolyn, and Jacquelin Burgess. 1994. "Social Constructions of Nature: A Case Study of Conflicts Over the Development of Rainham Marshes." *Transactions of British Geographers* 19: 291–310.

Hartzog, George, Jr. 1988. *Battling for the National Parks.* Mt. Kisco, N.Y.: Moyer Bell.

Heller, Alfred, ed. 1972. *The California Tomorrow Plan.* Los Altos, Calif.: Kaufmann.

Hines, Thomas. 1974. *Daniel Burnham.* New York: Oxford University Press.

Hogan, Michael, ed. 1992. *The End of the Cold War: Its Meaning and Implications.* New York: Cambridge University Press.

Hornbeck, David. 1990. "Spanish Legacy in the Borderlands." In *The Making of the American Landscape,* edited by Michael Conzen, 51–62. Boston: Unwin Hyman.

Huth, Hans. 1957. *Nature and the American: Three Centuries of Changing Attitudes.* Berkeley: University of California Press.

Jacobs, Allen. 1983. "1968: Getting Going, Staffing Up, Responding to Issues." In *Introduction to Planning History in the United States,* edited by Donald Krueckeberg, 235–257. New Brunswick, N.J.: Center for Urban Policy Research.

Jacobs, John. 1994, May 7. "Saving the Presidio: Politics in Review." *Sacramento Bee.*

Jakle, John. 1990. "Landscapes Redesigned for the Automobile." In *The Making of the American Landscape,* edited by Michael Conzen, 293–310. Boston: Unwin Hyman.

Jezierski, Louise. 1990. "Neighborhoods and Public-Private Partnerships in Pittsburgh." *Urban Affairs Quarterly* 26(2): 217–249.

Johnston, Chalmers, Laura D'Andrea Tyson, and John Zysman, eds. 1989. *Politics and Productivity: The Real Story of Why Japan Works.* Cambridge, Mass.: Ballinger.

Johnston, R. J., Peter Taylor, and Michael Watts, eds. 1995. *Geographies of Global Change: Remapping the World in the Late 20th Century.* Cambridge, Mass.: Blackwell.

Katz, Cindi, and Andrew Kirby. 1991. "In the Nature of Things: The Environment and Everyday Life." *Transactions of British Geographers* 16: 259–271.

Kennedy, Paul. 1987. *The Rise and Fall of the Great Powers: Economic Change and Military Conflict from 1500–2000.* New York: Vintage Books.

Kirby, Andrew, ed. 1992. *The Pentagon and the Cities.* Newbury Park, Calif.: Sage.

Kroeber, Alfred. 1925. *Handbook of the Indians of California.* Washington, D.C.: GPO.

Krueckeberg, Donald. 1983. *Introduction to Planning History in the United States.* New Brunswick, N.J.: Center for Urban Policy Research.

Lall, Betty, and John Tepper Marlin. 1992. *Building a Peace Economy: Opportunities and Problems of Post–Cold War Defense Cuts.* Boulder, Colo.: Westview Press.

Lawrence, Christine. 1988, July 16. "House Modifies, Passes Base-Closing Bill." *Congressional Quarterly,* 1976.

Liebert, Larry. 1989, February 23. "Navy Has Its Sights on Presidio Houses." *San Francisco Chronicle,* A2.

Lowell, John Bean, ed. 1994. *The Ohlone Past and Present: Native Americans of the San Francisco Bay Region.* Menlo Park, Calif.: Ballena Press.

Lufkin, Liz. 1989, February 17. "About-Face Ideas for the Presidio." *San Francisco Chronicle,* B3.

Lynch, Allen. 1992. *The Cold War Is Over—Again.* Boulder, Colo.: Westview Press.

Lynch, John E. 1970. *Local Economic Development After Military Base Closures.* New York: Praeger.

Malecki, Edward J. 1984. "Military Spending and the U.S. Defense Industry: Regional Patterns of Military Contracts and Subcontracts." *Environment and Planning C: Government and Policy* 2: 31–44.

Margolin, Malcolm. 1978. *The Ohlone Way: Indian Life in the San Francisco–Monterey Bay Area.* Berkeley, Calif.: Heyday Books.

Marin Independent Journal. 1994, May 11. "Promise, Peril Apparent as Congress Debates Presidio," A1, A7.

Markusen, Ann, Peter Hall, Scott Campbell, and Sabina Deitrick. 1991. *The Rise of the Gunbelt: The Military Remapping of Industrial America.* New York: Oxford University Press.

Markusen, Ann, and Joel Yudken. 1992. *Dismantling the Cold War Economy.* New York: Basic Books.

Massey, Doreen. 1994. *Space, Place and Gender.* Minneapolis: University of Minneapolis Press.

Matier, Phillip, and Andrew Ross. 1993, November 22. "Army Wants to Stay the (Golf) Course." *San Francisco Chronicle,* A15.

McCabe, Michael. 1992, May 9. "Gorbachev Declares Presidio Perfect Site for His Foundation." *San Francisco Chronicle,* A1.

McCoy, Charles. 1994, April 19. "Astonishing Views, and Many Opinions; Must Be the Presidio." *Wall Street Journal,* A1.

Meinig, Donald W., ed. 1979. *The Interpretation of Ordinary Landscapes.* New York: Oxford University Press.
Melman, Seymour. 1988. *The Demilitarized Society: Disarmament and Conversion.* Montreal: Harvest House.
Melman, Seymour. 1988, May. "Economic Consequences of the Arms Race: The Second-Rate Economy." *American Economic Review,* 55–59.
Melman, Seymour. 1986. "Swords into Ploughshares." *Technology Review* 89(1): 37–45.
Melman, Seymour. 1985. *The Permanent War Economy.* New York: Simon and Schuster.
Melman, Seymour. 1973. *Planning for Conversion of Military-Industrial and Military Base Facilities.* Washington, D.C.: U.S. Department of Commerce, Economic Administration, Office of Technical Assistance.
Melosi, Martin. 1993. "The Place of the City in Environmental History." *Environmental History Review* 17(1): 1–23.
Melosi, Martin. 1981. *Garbage in the Cities: Refuse, Reform and the Environment, 1880–1980.* College Station: Texas A&M University Press.
Melosi, Martin, ed. 1980. *Pollution and Reform in American Cities, 1870–1930.* Austin: University of Texas Press.
Merchant, Carolyn. 1980. *The Death of Nature: Women, Ecology and the Scientific Revolution.* London: Wildwood House.
Mills, C. Wright. 1956. *The Power Elite.* New York: Oxford University Press.
Mills, Mike. 1988, December 31. "Base Closings: The Political Pain Is Limited." *Congressional Quarterly,* 3625–3629.
Mills, Mike. 1988, July 2. "Members Go on the Offensive to Defend Bases." *Congressional Quarterly,* 1815–1817.
Mosely, Hugh G. 1985. *The Arms Race: Economic and Social Consequences.* Lexington, Mass.: Heath.
Muller, Edward K. 1990. "The Americanization of the City." In *The Making of the American Landscape,* edited by Michael Conzen, 269–293. Boston: Unwin Hyman.
Mumford, Lewis. 1961. *The City in History: Its Origins, Its Transformations, and Its Prospects.* New York: Harcourt Brace Jovanovich.
Nash, Roderick. 1982. *Wilderness and the American Mind.* New Haven, Conn.: Yale University Press.
National Parks and Conservation. 1994, January–February. Editorial, "The Presidio Update," 10.
New Yorker. 1995, April 22. "The Talk of the Town," 37–38.
Newsweek. 1989, November 27. "The Russians Aren't Coming," 48–50.
Nicholson-Lord, David. 1987. *The Greening of the Cities.* London: Routledge & Kegan Paul.
Nijkamp, Peter, and A. Perrels. 1994. *Sustainable Cities in Europe.* London: Earthscam.

Nolte, Carl. 1994, October 7. "Presidio Soldiers Furl Their Colors." *San Francisco Chronicle*, A1, A17.

Nolte, Carl. 1994, March 10. "Congressman Tries to Block Presidio Plan." *San Francisco Chronicle*, A1.

Nolte, Carl. 1994, January 6. "Army Trying to Keep Presidio Golf Course." *San Francisco Chronicle*, A1, A13.

Nolte, Carl. 1993, May 31. "Mount Tamalpais Eyesore Is Bureaucratic Legacy." *San Francisco Chronicle*, A13.

Nolte, Carl. 1991, November 15. *San Francisco Chronicle*, A25.

Oelschlaeger, Max. 1991. *The Idea of Wilderness: From Prehistory to the Age of Ecology.* New Haven, Conn.: Yale University Press.

OhUallachain, Breandan. 1987. "Regional and Technological Implications of the Recent Buildup in American Defense Spending." *Annals of the Association of American Geographers* 77: 208–223.

Paddock, Richard. 1993, October 1. "Global Center Plan Unveiled for Presidio." *Los Angeles Times*, A1.

Paddock, Richard. 1992, August 12. "View from Presidio: Profits." *Los Angeles Times*, A1.

Palmer, Elizabeth. 1991, July 6. "Commission Spares Just a Few from Cheney's Hit List." *Congressional Quarterly*, 1845–1847.

Palmer, Elizabeth. 1991, April 20. "Commission Comes to Life, Vowing a Fresh Look." *Congressional Quarterly*, 994–997.

Palmer, Elizabeth, and Pat Towell. 1991, April 13. "Cheney Reveals New Hit List; Members Feel the Pain." *Congressional Quarterly*, 931–932.

Paloú, Fray Francisco. 1913. *Life and Apostolic Labors of the Venerable Father Junipero Serra.* Pasadena, Calif.: G. W. James.

Penfold, Joseph, Stanley Cain, Richard Estep, Brock Evans, Roderick Nash, Douglas Schwartz, and Patricia Young. 1972. "Preservation of National Park Values: Report of a Task Force Assembled by the Conservation Foundation." In *National Parks for the Future: An Appraisal of the National Parks as They Begin Their Second Century in a Changing America*, 31–46. Washington, D.C.: Conservation Foundation.

Petulla, Joseph. 1988. *American Environmental History.* 2nd ed. Columbus, Ohio: Merrill.

Petulla, Joseph. 1977. *American Environmental History: The Exploitation and Conservation of Natural Resources.* San Francisco: Boyd and Fraser.

Polak, Fred. 1961. *The Image of the Future: Enlightening the Past, Orienting the Present, Forecasting the Future.* 2 vols. New York: Oceanna Publications.

Pursell, Carroll, Jr. 1972. *The Military-Industrial Complex.* New York: Harper and Row.

Rapp, David. 1988, February 20. "Deficit Limits Reagan's Options in 1989 Budget." *Congressional Quarterly*, 327–331.

Relph, Edward. 1976. *Place and Placelessness.* London: Pion.

Rees, John, Geoffrey Hewings, and Harold Stafford, eds. 1981. *Industrial Location and Regional Systems: Spatial Organization in the Economic Sector.* New York: J. F. Bergin.

Rettie, Dwight F. 1995. *Our National Park System: Caring for America's Greatest Natural and Historic Treasures.* Urbana: University of Illinois Press.

Ridenour, James M. 1994. *The National Parks Compromised: Pork Barrel Politics and America's Treasures.* Merriville, Ind.: ICS Books.

Rockefeller, Laurance. 1994. "A Report to the President and to Congress by the Outdoor Recreation Resources Review Commission, Laurance S. Rockefeller, Chairman, January 1962." In *America's National Park System: The Critical Documents,* edited by Lary Dilsaver, 224–236. Lanham, Md.: Rowman and Littlefield.

Rose, Steven. 1973. *Testing the Theory of the Military-Industrial Complex.* Lexington, Mass.: Heath.

Rothman, Hal. 1989. *Preserving Different Pasts: The American National Monuments.* Urbana: University of Illinois Press.

Rubenstein, Steve. 1996, May 31. "Readers Far and Wide Overwhelm Caen with Good Wishes." *San Francisco Chronicle,* E1.

Runte, Alfred. 1987. *National Parks: The American Experience.* 2nd ed. Lincoln: University of Nebraska Press.

Russell, Sabin. 1994, April 23. "Wilson Makes Plea in S.F. for Presidio Park." *San Francisco Chronicle,* A17.

Sandalow, Marc. 1994, June 30. "GOP Backs Off—Presidio Plan Gets Preliminary OK." *San Francisco Chronicle,* A3.

Sandalow, Marc. 1994, February 1. "Babbitt Wearily Confident About Future of Presidio." *San Francisco Chronicle,* A3.

Sandalow, Marc. 1993, October 27. "House GOP Assails Cost of Presidio Plan." *San Francisco Chronicle,* A1.

San Francisco Bay Guardian. 1994, January 12. Editorial, "The Presidio in Peril."

San Francisco Bay Guardian. 1993, December 29. Editorial.

San Francisco Chronicle. 1994, January 6. Editorial, "The Army Role as a Tenant at the Presidio."

San Francisco Examiner. 1994, May 8. Editorial, "Don't Abandon the Presidio," A14.

San Francisco Examiner. 1994, April 19. Editorial, "Pelosi to Amend Plans for Presidio," A6.

San Francisco Examiner. 1993, August 26. Editorial, "Paying for the Presidio," A22.

San Francisco Examiner. 1989, August 13. Editorial, "Presidio: It's in Our Hands," A22.

San Francisco Independent. 1994, May 3. Editorial, "Pelosi's Presidio Solution."

Sawers, Larry, and William K. Tabb, ed. 1984. *Sunbelt/Snowbelt: Urban Development and Regional Restructuring.* New York: Oxford University Press.

Scharma, Simon. 1995. *Landscape and Memory.* New York: Knopf.

Schmitt, Peter. 1990. *Back to Nature: The Arcadian Myth in Urban America.* Baltimore: Johns Hopkins University Press.

Scott, Mel. 1985. *The San Francisco Bay Area: A Metropolis in Perspective*. 2nd ed. Berkeley: University of California Press.

Scott, Stanley, ed. 1966. *San Francisco Bay Area: Its Problems and Future*. Berkeley Institute of Governmental Studies. Berkeley: University of California Press.

Seth, Vikram. 1986. *The Golden Gate, a Novel in Verse*. New York: Vintage Books.

Shelley, Kevin. 1994, March 24. "The Plan to Sell the Presidio." *San Francisco Examiner*, A21.

Short, John Rennie. 1991. *Imagined Country: Society, Culture and Environment*. London: Routledge.

Short, John Rennie. 1989. *The Humane City: Cities as If People Really Mattered*. Oxford: Basil Blackwell.

Smith, Herbert H. 1961. *The Citizen's Guide to Planning*. West Trenton, N.J.: Chandler-Davis.

Snow, Donald. 1991. *The Shape of the Future: The Post–Cold War World*. Armonk, N.Y.: M. E. Sharpe.

Soulé, Michael, and Gary Lease, eds. 1995. *Reinventing Nature? Responses to Postmodern Deconstruction*. Washington, D.C.: Island Press.

Specht, Jim. 1994, May 13. "Senators Worry Whether Presidio Will Drain Budget." *Marin Independent Journal*, B5.

Spence, Mary Lee, and Donald Jackson, eds. 1973. *The Expedition of John C. Frémont*. Urbana: University of Illinois Press.

Spolar, Christine. 1993, December 28. "Designs of Park Service, Pentagon Differ on Converting the Presidio." *Washington Post*.

Starr, Kevin. 1988, November 20. "The Presidio." *San Francisco Examiner, Image*, 72–76.

Starr, Kevin. 1973. *Americans and the California Dream, 1850–1915*. Santa Barbara, Calif.: Peregrine Smith.

Stein, Loren. 1995, October 10. "Congress Rides to the Presidio's Rescue." *Christian Science Monitor*, 10–11.

Steinstra, Tom. 1994, April 10. "Presidio Park Plan Stirring Heated Debate." *San Francisco Examiner*, D12.

Stephenson, Max, Jr. 1991. "Whither the Public-Private Partnership: A Critical Overview." *Urban Affairs Quarterly* 27(1): 109–127.

Taft, William. 1987, October 5. Editorial, "When Doves Turn into Hawks." *Washington Times*, D1.

Talbot, Strobe. 1989, December 18. "Reciprocity at Last." *Time*, 40.

Temko, Allen. 1989, May 15. "After 200 Years, the Presidio Will Get Its First Master Plan." *San Francisco Chronicle*, A6.

Temko, Allen. 1989, May 15. "Time Is Now to Plan a Visionary Future." *San Francisco Chronicle*, A1.

Thomas, Keith. 1983. *Man and the Natural World*. London: Allen Lane.

Time. 1990, January 1. "Rethinking the Red Menace."

Tira, Peter. 1993, December 14. "Area's Military Vets Go to Battle over Presidio Hospital." *San Francisco Independent*, A1.

Tocqueville, Alexis de. 1966. *Democracy in America*. 2 vols. New York: Harper and Row.

Towell, Pat. 1993, March 20. "Tough Calls Are Put on Hold; Clinton Keeps Options Open." *Congressional Quarterly*, 678.

Towell, Pat. 1988, October 15. "Hill Paves Way for Closing Old Bases." *Congressional Quarterly*, 2999.

Towell, Pat. 1988, September 10. "Tough Challenges Await the 41st President." *Congressional Quarterly*, 2491–2496.

Towell, Pat. 1988, February 27. "The Carlucci Budget: Picking Among Priorities." *Congressional Quarterly*, 522–527.

Trubowitz, Peter, and Brian Roberts. 1991. *Bearing the Burden: Distributive Politics and National Security*. Working Paper, John M. Olin Institute for Strategic Studies, Harvard University.

Tuan, Yi-Fu. 1974. *Topophilia: A Study of Environmental Perception, Attitude and Values*. Englewood Cliffs, N.J.: Prentice-Hall.

U.S. News & World Report. 1989, December 11. Cover page, "After the Cold War: Do We Need an Army?"

Ventresca, Joel. 1994, January 18. "Keep the Presidio Ours." *San Francisco Independent*.

Walker, Thaai. 1994, April 25. "1,500 March in Support of Presidio Park." *San Francisco Chronicle*, A15.

Wall Street Journal. 1989, August 28. "Letters to the Editor," A9.

Walsh, Diana. 1989, February 12. "The Presidio Inspires Me to Dream." *San Francisco Examiner*, B1.

Ward, David. 1971. *Cities and Immigrants: A Geography of Change in Nineteenth Century America*. New York: Oxford University Press.

Wellman, J. Douglas. 1987. *Wildland Recreation Policy: An Introduction*. Malabar, Fla.: Krieger Publishing.

White, M., and L. White. 1962. *The Intellectual Versus the City*. Cambridge, Mass.: Harvard University Press and MIT Press.

Whittlesey, Derwent. 1929. "Sequent Occupance." *Annals of the Association of American Geographers* 19: 162–166.

Wilbur, Marguerite Eyer, ed. 1954. *Vancouver in California, 1792–1794: The Original Account of Vancouver*. 3 vols. Los Angeles: Glen Dawson.

Wilbur, Marguerite Eyer, ed. 1937. *Duflot De Mofra's Travels on the Pacific Coast*. 2 vols. Santa Ana, Calif.: Fine Arts Press.

Wilson, Alexander. 1992. *The Culture of Nature: North American Landscape from Disney to the Exxon Valdez*. Cambridge, Mass.: Blackwell.

Wilson, William H. 1990. *The City Beautiful Movement*. Baltimore: Johns Hopkins University Press.

Wilson, William H. 1983. "Moles and Skylarks." In *Introduction to Planning History in the United States*, edited by Donald Krueckeberg, 88–121. New Brunswick, N.J.: Center for Urban Policy Research.
Wirth, Conrad. 1980. *Parks, Politics and the People*. Norman: University of Oklahoma Press.
Work, Hubert. 1994. "Statement of National Park Policy—Secretary Work, March 11, 1925." In *America's National Park System: The Critical Documents*, edited by Lary Dilsaver, 62–65. Lanham, Md.: Rowman and Littlefield.
Worster, Donald. 1977. *Nature's Economy: A History of Ecological Ideas*. Cambridge: Cambridge University Press.
Wright, Will. 1992. *Wild Knowledge: Science, Language, and Social Life in a Fragile Environment*. Minneapolis: University of Minneapolis Press.
Yergin, Daniel. 1977. *Shattered Peace: The Origins of the Cold War and the National Security State*. Boston: Houghton Mifflin.
Zelinksy, Wilbur. 1990. "The Imprint of Central Authority." In *The Making of the American Landscape*, edited by Michael Conzen, 335–354. Boston: Unwin Hyman.

Public and Personal Documents

Boxer, Barbara. 1994, May 12. "Statement of Senator Barbara Boxer, Senate Energy and Natural Resources Subcommittee on Public Lands, National Parks, and Forests (S. 1639)." *News from U.S. Senator Barbara Boxer*.
Congressional Record for the House of Representatives. 1993, July 15. "Debate on Duncan Amendment to H.R. 2520 FY 94 Interior Appropriations Bill."
Congress of the United States. House of Representatives. Committee on Armed Services. 1989. *Report of the Defense Secretary's Commission on Base Realignment and Closure Hearings Before the Military Installations and Facilities Subcommittee, February 22 and March 1, 1989*. Washington, D.C.: GPO.
Congress of the United States. House of Representatives. Committee on Armed Services. 1988. *Hearing on Base Closure: Military Installations and Facilities Subcommittee Hearing on H.R. 1583 and Military Installations and Facilities Subcommittee and Defense Policy Panel Joint Hearing on H.R. 4481, March 17, 18, 19, and June 8, 1988*. Washington, D.C.: GPO.
Congress of the United States. House of Representatives. Committee on Banking, Finance, and Urban Affairs. 1988. *Economic Conversion: Hearing Before the Subcommittee on Economic Stabilization, June 29, 1988*. Washington, D.C.: GPO.
Congress of the United States. House of Representatives. Committee on Interior and Insular Affairs. Subcommittee on National Parks and Recreation. 1972. *Hearings on H.R. 9498 to Establish a National Recreation Area in San Francisco and Marin Counties, August 9 and May 11 and 12, 1972*. Washington, D.C.: GPO.
Congress of the United States. Office of Technology Assessment. 1992, February. *After the Cold War: Living with Lower Defense Spending*. Washington, D.C.: GPO.

Congress of the United States. Senate. 1994. *Management of the Presidio in San Francisco by the National Park Service, Hearing Before the Subcommittee on Public Lands, National Parks, and Forests of the Committee on Energy and Natural Resources, May 12, 1994*. Washington, D.C.: GPO.

Congress of the United States. Senate. 1993. *Management of the Presidio in San Francisco by the National Park Service, Hearing Before the Subcommittee on Public Lands, National Parks, and Forests of the Committee on Energy and Natural Resources, October 25 and 26, 1993*. Washington, D.C.: GPO.

Draft Letter by Presidio Council for Pete Wilson to Send to Congressman Bruce Vento, Chair, House Subcommittee on National Parks, Forests, and Public Lands. 1993. Document courtesy of William K. Reilly.

Duncan, John. 1993, July 14. Letter to Congress, House of Representatives. July 14, 1993. Document courtesy of the Golden Gate National Parks Association.

Feinstein, Dianne. 1994, May 12. "Statement by Senator Dianne Feinstein, Hearing on Presidio Legislation (S. 1549 and S. 1639), Subcommittee on Public Lands, National Parks, and Forests." Document obtained at the Senate hearing.

Golden Gate National Park Association. 1996, Winter. *Gateways: Quarterly Newsletter*. San Francisco: GGNPA.

Golden Gate National Park Association. 1994. *Presidio Update: An Update of the National Park Service Presidio Transition Process*. San Francisco: GGNPA.

Golden Gate National Park Association. 1994, Summer. *Presidio Update: An Update of the National Park Service Presidio Planning Process*. San Francisco: GGNPA.

Golden Gate National Park Association. 1992, October 3. *Presidio Update: An Update of the National Park Service Presidio Planning Process*. San Francisco: GGNPA.

Golden Gate National Park Association. 1990, October 10. Memo, "The Presidio Task Force." Document courtesy of William K. Reilly.

Kennedy, Roger. 1993, October 8. Press release, "Preamble to the Presidio Plan." U.S. Department of the Interior.

Keyser, Jerry. 1993, October 26. "Statement of Jerry Keyser, Keyser Marston Associates," to the House Subcommittee on National Parks, Forests, and Public Lands. Document obtained at the hearing.

Langelier, John Phillip, and Daniel B. Rosen. 1992. *Historic Resource Study: El Presidio de San Francisco, a History Under Spain and Mexico, 1776–1846*. Denver: U.S. Department of the Interior, National Park Service.

Middleton, Craig. Presidio Council. 1993, October 29. Memo to William K. Reilly. "D.C. Trip Report." Document courtesy of William K. Reilly.

National Park Service. 1993, October. *Creating a Park for the 21st Century: From Military Post to National Park*. Draft General Management Plan Amendment, Presidio of San Francisco. NPS D-148. San Francisco: GGNRA.

National Park Service. 1993, October. *Draft General Management Plan Amendment: Environmental Impact Statement, Presidio of San Francisco*. NPS D-149. San Francisco: GGNRA.

National Park Service. 1993. *Creating a Park for the 21st Century: From Military Post to National Park*. San Francisco: National Park Service, Presidio Project Office.

National Park Service. 1993. *National Parks for the 21st Century: The Vail Agenda*. Post Mills, Vt.: Chelsea Green Publishing Co. / National Park Service.

National Park Service. 1991, Winter. *Presidio Visions Kit*. San Francisco: GGNPA and GGNRA.

National Park Service. 1985. *Presidio of San Francisco National Historic Landmark District: Historic American Buildings Survey Report*. San Francisco: U.S. Department of the Interior and U.S. Department of Defense.

Pelosi, Nancy. 1993, November. *Presidio Report from Congresswoman Nancy Pelosi, U.S. House of Representatives*.

Pelosi, Nancy. 1993, November 3. "Pelosi Introduces Long-Term Presidio Bill." *News from Congresswoman Nancy Pelosi*.

Pelosi, Nancy. 1993, July 15. "Attempt to Slash Presidio Funding Fails." *News from Congresswoman Nancy Pelosi*.

Pelosi, Nancy, Dianne Feinstein, and Barbara Boxer. 1993. Letter to President Clinton. April, 14, 1993. Document courtesy of William K. Reilly.

People for Open Space. 1983. *Room Enough: Housing and Open Space in the Bay Area*. San Francisco: People for Open Space.

Public Law 84-825. 1970. "An Act to Improve the Administration of the National Park System by the Secretary of the Interior, and to Clarify the Authorities Applicable to the System, and for Other Purposes."

Reilly, William K. 1993, October 26. "Statement of William K. Reilly," to the House Subcommittee on National Parks, Forests, and Public Lands. Document obtained at the hearing.

Rosenblatt, Toby. 1993, October 25. "Statement of Toby Rosenblatt, Vice Chair, Presidio Council, Chair, Golden Gate National Park Association," to the House Subcommittee on National Parks, Forests, and Public Lands. Document obtained at the hearing.

San Francisco Bay Conservation and Development Commission. 1969. *San Francisco Bay Plan*.

Thompson, Erwin, and Sally Woodbridge. 1992. *Special History Study: Presidio of San Francisco, an Outline of Its Evolution as a U.S. Army Post, 1847–1990*. Denver: U.S. Department of the Interior, National Park Service.

U.S. Arms Control and Disarmament Agency. 1970. *Economic Impact of Military Base Closings: Adjustments by Communities and Workers*. Vol. 1. Washington, D.C.: GPO.

U.S. Department of Commerce. Office of Area Development. 1959. *The Future Development of the San Francisco Bay Area, 1960–2020*. Washington, D.C.: GPO.

U.S. Department of Defense. Base Closure and Realignment Commission. 1995, March. *Base Closure and Realignment Commission Report, 1995*. Washington, D.C.: GPO.

U.S. Department of Defense. Defense Conversion Commission. 1992, December 31. *Adjusting to the Drawdown: Report of the Defense Conversion Commission.* Washington, D.C.: GPO.

U.S. Department of Defense. Defense Secretary's Commission on Base Realignment and Closure. 1988, December. *Report of the Defense Secretary's Commission.* Washington, D.C.: GPO.

U.S. Department of Defense. President's Economic Adjustment Committee. 1986. *25 Years of Civilian Re-use, 1961–1986: A Summary of Completed Military Base Economic Adjustment Projects.* Washington, D.C.: The Pentagon.

U.S. Department of the Interior. National Park Service. 1994, June 21. News release, "National Park Service Selects Tides Foundation and UCSF for Letterman Talks." Document courtesy of the Golden Gate National Park Association.

U.S. Department of the Interior. 1994, March 14. Memo, "Draft Comments on H.R. 3433." Document courtesy of William K. Reilly.

U.S. General Accounting Office. 1993, October 26. *Report on the Transfer of the Presidio from the Army to the National Park Service.* Testimony before the House Subcommittee on National Parks, Forests, and Public Lands.

Interviews

Chandler, Robert. National Park Service, Presidio General Manager. Interview. San Francisco, May 31, 1996.

Dilsaver, Lary. Professor of Geography, University of South Alabama. An informal discussion at the Annual Association of American Geographers Conference. Fort Worth, Tex., April 1997.

Huse, Brian. Director, Pacific Regional Office, National Parks and Conservation Association. Interview. Oakland, Calif., May 28, 1996.

Lemons, Judith. Administrative Assistant to Congresswoman Nancy Pelosi. Washington, D.C., July 25, 1997.

Middleton, Craig. Director of Government Relations, Golden Gate National Park Association. Interview. San Francisco, May 28, 1996.

Moore, Greg. Executive Director, Golden Gate National Park Association. Interview. San Francisco, May 23, 1996.

Neubacher, Don. Superintendent, Point Reyes National Seashore (and former Presidio planning team captain). Interview. Point Reyes, Calif., May 25, 1996.

O'Neill, Brian. Superintendent, Golden Gate National Recreation Area, National Park Service. Interview. San Francisco, May 28, 1996.

Reilly, William K. Senior Adviser to the Presidio Council. Telephone interview. April 5, 1996.

Reilly, William K. Telephone interview. December 12, 1995.

Reilly, William K. Discussions and informal interviews. June 1993–March 1994.

Rosenblatt, Toby. Chair, Golden Gate National Park Association Board of Directors. Interview. San Francisco, May 28, 1996.

Schwab, Michael. Graphic artist. Recording of Remarks Made at "From Park to Art: Meet the Artist" (an event sponsored by GGNPA), Presidio Visitor's Center, Building 102, Main Parade Ground, San Francisco, June 1, 1996.

INDEX

adobe buildings, 23–24, 40
Ailes, Stephen, 52
air pollution, 54
Alaskan Wilderness, 194, 248 n. 61
Albright, Horace, 149, 157
Alcatraz, 22–23, 25, 57, 60, 228 n. 7
Alexander, Michael, 124, 133
Alioto, Joseph, 64
Alta California, 7, 11, 14, 216 n. 6
American Institute of Architects, 137
America's National Parks and Their Keepers (Foresta), 173
Angel Island, 22–23, 50, 60
antiurbanism, 246 n. 35
Anza, Juan Bautista de, 8
Aquatic Park, 57
architectural styles: of military bases, 31, 219 n. 39; Mission Revival style, 42, 44 fig. 4.3, 62, 102, 105; officers' quarters, 27 fig. 3.4; of Presidio, 62, 63; Spanish heritage of, 42–44, 48
Armey, Dick, 75, 79–80, 226 n. 51
Army. *See* U.S. Army
Arrigo, Joseph, 228 n. 4
Arts and Craft movement, 220 n. 5
Aspin, Les, 82, 118, 226–27 n. 58
Association for Improvement and Adornment of San Francisco, 37
automobile boom, 46–47
aviation technology museum, 104
Ayala, Juan Manuel de, 8

Babbitt, Bruce, 119, 130
Baker Beach, 62, 106
Baltimore (Maryland), 199
Base Closure and Realignment Commission (BRAC), 75–76, 78, 80–85, 117, 226 n. 54, 226 n. 56
battlefield sites, 154, 157
Bay Bridge, 49
beautification programs, 30–33, 35–39, 40, 63, 198
Benicia, 22
Berlin Wall, 76–77
Blakeley, Edward, 143
Boston (Massachusetts), 199
boundary proposals, 22, 23 fig. 3.1, 132
bowling alley, 234 n. 3
Boxer, Barbara, 82, 87–88, 111, 125, 135, 137–38, 141
buildings, at Presidio: on Crissy field, 104; current, 4; earthquake damage to, 40; first, 13; improvements to, 23–24, 30–31; need for additional, 42; Officers' Club, 40, 48; officers' quarters, 27 fig. 3.4; remodeling of, 48; square footage of, 62
Bumpers, Dale, 137
Burnham, Daniel Hudson, 37–38, 39, 40, 42, 179, 230 n. 16
Burton, Philip, 57, 59, 60–61, 131, 151, 167, 174, 179
Burton legislation, 88, 118, 181
Bush, George, 75, 76, 83, 128, 226 n. 56, 233–34 n. 2

271

Cabrillo, Juan, 7
Caen, Herb, 53, 221–22 n. 39
California, acquisition of, 20
California Desert Protection Act, 242–43 n. 88
California Tomorrow task force, 54
Call for Interest proposal, 99–100, 232 n. 43
Cape Cod National Seashore, 159–60
Capron, Elisha Smith, 25
Carlucci, Frank, 75, 86
Catholicism, Indians' conversion to, 10
cavalry stables, 105
cemeteries, 105, 128, 157, 240 n. 53
Central Park (New York), 198
Chamber of Commerce, 137
Chandler, Robert, 141, 189, 235 n. 20, 237–38 n. 34
Chase, Alston, 149
Cheney, Richard, 77–78, 82, 226 n. 54
cities: greening of, 198–200; and parks, 197–98
Citizens for Regional Recreation and Parks, 55
City Beautiful movement, 35–39, 40, 198
civic improvement societies, 37
Civil War, 26, 154
Civilian Conservation Corps (CCC), 48, 157
Civilian Works Administration (CWA), 48
Clean Air Act, 169
Clean Water Act, 169
Cleveland (Ohio), 199
Cliff House, 57
Clinton, Bill, 83–84, 125, 143, 237–38 n. 34
coastal bluffs, 106
Cold War, end of, 76–78
communism, collapse in Eastern Europe, 76–77
community mobilization, 136–37
community service, 98, 99 table 6.2, 102
conference center, 105
Congress (*see also* legislation; politics): accountability of Presidio Trust, 142, 189; control of parks, 185–86; legislative process, 243 n. 90, 244 n. 99; opposition to Grand Vision, 127–38; park politics in, 182–85; park system agenda of, 187–90
Conservation Foundation, 161–62, 247 n. 40, 248 n. 58
conservation movement, 151–52
constituency, building, 187–90
contingency plans, 108–10
conversion planning: funding of, 92; initiation of, 86–88; National Park Service mobilization, 92–94; production phase of, 98–100; public participation in, 93–94; scope of, 94–96; suggestions for, 88–90; timeline for, 97 fig. 6.1; visioning phase of, 95–96, 98
Coon, Jim, 129
Corrigan, Robert, 89
costs (*see also* funding): of alternative plans, 110 table 6.4; of Grand Vision, 106, 107 table 6.3, 134, 183–84; of maintenance, 135; of military base closures, 84 table 5.1; of Presidio, 134 table 7.2
Cranston, Alan, 56
Crissy field, 45–46, 46 fig. 4.4, 51, 62, 89, 104, 108–9, 116
cross-cultural cooperation, 98, 99 table 6.2
crown jewel syndrome, 162–63, 165–67, 172–73, 187
cultural landscape, 38–39, 220 n. 61
culture parks, 157–58, 164 fig. 8.2, 165, 196

Dana, Richard Henry, 18
Defense Base Closure and Realignment Act, 67, 80, 83, 116
defense spending: and base closures, 84, 227 n. 65; cycles of, 67–70, 68 fig. 5.1; during downsizing, 234 n. 3; as percentage of state purchases, 72 fig. 5.3; in post–Cold War era, 77–78; on Presidio, 115, 242 n. 80; during Reagan buildup, 73–76; spatial consequences of, 70–73; terminology of, 223 n. 2
Denver (Colorado), 198
Depression, 47–48
development projects, threats from, 168–69
Devol, C. A., 43
Dilsaver, Lary, 149, 247 n. 47
DiMaggio, Joe, 235 n. 13
Dole, Robert (Bob), 142, 143, 242–43 n. 88
Drake, Francis, 7
dualisms, 191–92
Duncan, John, 128–35, 136, 145, 169, 184, 241 n. 71

earthquake, 40, 41 fig. 4.1
East Housing Area, 105
economic adjustment, 70
economic conversion, 70, 223 n. 4
economic prosperity, 53
educational centers, 102, 103, 105, 146
Eisenhower, Dwight D., 70
El Camino Del Mar, 47
El Camino Real, 11
Emerson, Ralph Waldo, 27
Endangered Species Act, 169

endangered species list, 62
England, 15
environmental contamination, 115–16, 227 n. 2, 234 n. 6, 234 n. 7, 234 n. 8
Environmental Defense Fund, 137
environmental education center, 103, 146
environmental geography, 190–91, 251 n. 25
environmental-impact statements, 100
Environmental Protection Agency (EPA), 116–17
environmental regulations, 169, 227 n. 2
environmentalism: in land-use debates, 198–99; movements for, 26–28, 54–56, 218 n. 22
Espinoza, Martin, 111, 126, 127, 239–40 n. 49
Everglades National Park, 167–68
Expo, Panama-Pacific International, 45

federal budget deficit, 75–76, 128, 240 n. 51
federal planning requirements, 100
Feinstein, Dianne, 111, 125, 137–38, 141, 242–43 n. 88
filibusters, 138, 242–43 n. 88
fillmore, Millard, 22, 23 fig. 3.1, 26
fire control policy, 252–53 n. 43
fiscal conservatism, 128, 184
fisher, Donald, 143
fitzSimmons, Margaret, 251 n. 25
Font, Pedro, 9
Ford, Gerald, 79
Foresta, Ronald, 149, 173
Fort Baker, 50, 60
Fort Barry, 50
Fort McDowell, 50
Fort Mason, 50, 236–37 n. 25
Fort Point, 22–23, 25 fig. 3.3, 62, 104–5
Fort Scott, 42–44, 43 fig. 4.2, 44 fig. 4.3, 45, 52, 102
fortifications, 9, 22–26
Francisco de Ortega, José, 8
Fredericks, Steven, 116
Freemeyer, Allen, 140
Frémont, John, 20
Friends of the Earth, 55
funding (*see also* costs): clashes over, 128–31; of conversion, 104; of Grand Vision, 183–84; ongoing, 120, 123; of transition phase, 114–16
Funston, Frederick, 40, 42
future images, 181–82

Galvez, José de, 7–8
Garamendi, John, 143

Gateway National Recreation Area, 160, 222–23 n. 59
Geddes, Patrick, 180
Gejdenson, Sam, 80
General Authorities Act, 166–67
geography: discipline of, 251 n. 27; environmental, 190–91, 251 n. 25; historical, 4–5, 63–66, 215 n. 2; of Presidio, 3–4, 62
Gilliam, Harold, 33, 113, 199, 239–40 n. 49
Gingrich, Newt, 138, 142
Gold Rush, 20–23
Golden Gate, 20, 23, 24 fig. 3.2
Golden Gate, The (Seth), 64
Golden Gate Bridge, 48–50, 49 fig. 4.5, 66, 104–5
Golden Gate National Park Association (GGNPA), 61–62, 91–92, 147–49, 176–77, 179, 230 n. 17
Golden Gate National Recreation Area (GGNRA): establishment of, 57–63, 160, 179, 222–23 n. 59; incorporation into, 88–94; map of, 58 fig. 4.6; in planning process, 92–94, 95; Presidio's incorporation into, 86–87, 88, 90; support of, 222 n. 53; visitor center of, 102
Golden Gate Park, 29, 31, 32, 35, 198
golf course, 33, 62, 105, 118–19
Gorbachev, Mikhail, 74, 89, 229 n. 12
Gramm, Phil, 79
Gramm-Rudman-Hollings law, 75
Grand Canyon, 194
Grand Vision: alternative plans costs, 110 table 6.4; and community, 111–12, 179–81; concepts of, 98, 99 table 6.2; costs of, 106, 107 table 6.3, 134, 183–84; draft plan of, 100–106; inception of, 90; opposition to, 129–30; realization of, 178; support of, 136–37
Grant, Ulysses S., 152
grassland areas, 106
Great Britain, 15
Great Depression, 47–48
Greenpeace, 55, 199
Gunbelt region, 71–73, 71 fig. 5.2, 74, 224 n. 17

Hall, Peter, 180
Hansen, James, 130, 138
Hardie, James, 20
Hartzog, George, 149, 247–48 n. 52
Harvey, James (Jim), 91, 94, 121, 127, 230 n. 19, 237 n. 28
hazardous waste, 115–16, 227 n. 2, 234 n. 6, 234 n. 7, 234 n. 8

health and scientific discovery, 98, 99 table 6.2, 103–4
Hefley Bill, 141
Hetch Hetchy Valley, 168
Heyman, Ira Michael, 119
Hickel, Walter, 56
hiking trails, 104–5
historic landmark designations, 52, 57, 63
historical geography, 4–5, 63–66, 215 n. 2
historical preservation, 155, 157, 198
Historical Property Leasing Program, 232 n. 38
hospitals, 51, 105, 228 n. 4, 234 n. 3
Howard, Ebenezer, 180
human occupation, 6–7, 65
Hunter's Point, 50
Huth, Hans, 151

improvements, to Presidio, 23–24, 30–31, 50, 63
Indians: conflicts with, 24; conversion to Catholicism, 10; defenses against, 11; effects of European diseases, 13; preservation of historical sites, 155; in slavery, 18; uprisings of, 15, 16, 26
international cooperation, 98, 99 table 6.2

Jackson, Andrew, 19
Jackson Hole National Monument, 168–69
Johnson, Lyndon B., 69–70, 79
Jones, William Albert, 32
Jordan, Frank, 111, 137

Kennedy, Robert, 228 n. 7
Kennedy, Roger, 65–66, 137, 163, 179
Kennedy, Ted, 143
Kroeber, Alfred, 7

LAIR (Letterman Army Institute of Research), 103–4, 112
land speculation, 21–22, 52
landscape design, 31–33, 39
land-use politics, 198–99, 231 n. 25
legislation (*see also* Congress): clashes over, 127–38, 140–42, 182–85; of military base closures, 67; for Presidio Partnership, 123–25, 127, 129–30, 132–33, 136–37, 140–42; for Presidio Trust, 142–43, 144–45 fig. 7.4, 185, 242–43 n. 88, 244 n. 97, 244 n. 102, 244 n. 103
Lemons, Judith, 136
Letterman Complex, 103–4, 109, 237–38 n. 34
Letterman Hospital, 51, 228 n. 4, 234 n. 3

Levin, Carl, 80
lighthouses, 22
Lobos Creek, 106, 108–9
Lujan, Manuel, 94

McCarthy, Leo, 59
McKinsey & Co. study, 121, 236 n. 23
MacKintosh, Charles Rennie, 220 n. 5
McNamara, Robert, 69
Main Post, 102, 109
Mallory, Glynn, 119, 138
Manfield, Joseph, 23
Manifest Destiny, 18
Mare Island, 22, 50
Maritime Museum, 57
Marsh, George Perkins, 27, 151
marshes, 6, 31, 34, 45, 104
master plan, 95 (*see also* Grand Vision)
medical research, 103–4, 233 n. 65
Melman, Seymour, 70–71
Mexican-American War, 20
Mexico: cession of California, 20; declaration of independence, 17; occupation of Presidio, 17–19; Secularization Act, 18
Meyer, Amy, 57, 124, 143, 231 n. 24
Middleton, Craig, 186, 242 n. 80
military base closures: after Cold War, 224–25 n. 19; Congressional view of, 226–27 n. 58; costs and savings, 84 table 5.1; and federal deficit, 75–76; in Georgia, 227 n. 60; after Korean War, 69–70; legislation of, 67, 78–85; as peace dividend, 77–78; Presidio's inclusion in, 86–88; by state, 85 fig. 5.4
military garrisons, 9–12
military importance, of Presidio, 22, 30, 38–39, 46, 50–51, 52, 63
military-industrial complex, 70, 225 n. 27
Miller, George, 117, 129, 235 n. 11
Mission Dolores, 12 fig. 2.2, 13
Mission Revival architecture, 42, 44 fig. 4.3, 62, 102, 105
missions, 9–12
modernistic paradigm, 191
Mofras, Duflot de, 19
monumentalism, 152–53, 172–73, 187
Moore, Greg, 91, 126, 176–77, 185, 230 n. 18
Morris, William, 220 n. 5
Moses, Robert, 198
Mott, William Penn, 90, 96, 111, 230 n. 16
Mountain Lake, 8

mounted patrol museum, 105
Muir, John, 27, 168
Muir Woods, 57
Murphy, May, 143
museums, 57, 104, 105
mushroom gathering, 89, 229 n. 11

Nash, Roderick, 151
national cemeteries, 105, 157
National Environmental Policy Act, 95, 169
National Military Parks and Battlefield Act, 154
national monuments, 157
National Park Service: budget of, 135, 174, 233–34 n. 2; in conversion process, 98, 99–100; criticism of, 111, 185–86; establishment of, 156; legislative clashes over, 128–31, 133–34; mission of, 172; and nature/culture dualism, 195–97; negotiation with Army, 117–19; in planning process, 94–96, 100–101, 108–9; praise for, 186; Presidio's incorporation into, 88–94; in public-private partnership, 119–22; responsibilities of, 122 table 7.1; seventy-fifth anniversary of, 169–70; in transfer ceremony, 138–40; in transition phase, 114–17
National Park System, units of, 150 table 8.1
national parks: acreage of, 169, 248 n. 61; building constituency for, 187–90; commercial ventures in, 95, 232 n. 38; comparison of, 145–46, 146 table 7.3; as cultural expression, 152–54; decommissioning process, 252 n. 41; defining system, 155–56, 246 n. 19; dynamic evolution of, 170–72; entrance fees at, 247 n. 50; fire control policy, 252–53 n. 43; hierarchy of, 163, 164 fig. 8.2, 165–67, 171–72, 247–48 n. 52; infrastructure maintenance in, 174; public access to, 153, 246 n. 19; respect for, 186; shaping system of, 149–51; social creation of, 193–95; "thinning the blood" argument, 189, 249 n. 75; threats to, 168–69; "undeserving" parks, 173–76; visitor count, 169
National Parks: The American Experience (Runte), 152–53
National Parks and Recreation Act, 222–23 n. 59
national recreation areas, 158–62, 165–67, 172–73
Native Americans. *See* Indians
natural disasters, 13–14
nature, social construction of, 190–93
nature/culture dualism, 154, 187, 190–93, 195–97
nature parks, 157–58, 164 fig. 8.2, 165, 195–96

Neubacher, Don, 93
Nixon, Richard M., 56–57, 59, 160
Nunn, Sam, 227 n. 60

Ocean Beach, 57
Office of Economic Adjustment (OEA), 70
Officers' Club, 118 (*see also* buildings, at Presidio)
Ohlone Indians, 6–7, 13, 18, 216–17 n. 23
Olmsted, Frederick Law, 31–32, 37
Omnibus Parks and Public Lands Management Act (1996), 143, 244 n. 102
O'Neill, Brian, 93, 94, 110, 119, 130–31, 133, 175–76, 188
Operation Divot Storm, 117–19

Panama-Pacific International Exposition, 45
Park Closure Commission proposal, 140–41
park movements, 151–54
Park Partners, 100, 101
park plans, for Presidio (*see also* Presidio Plan): in nineteenth century, 28–30; political opposition to, 5; second wave of, 56–63
parkland acquisition, 55, 59–60
Parks to the People program, 57
Pelosi, Nancy: on base closures, 82; on Letterman complex lease, 237–38 n. 34; on military spending, 87–88; and Operation Divot Storm, 117, 119; for Presidio partnership, 123–25, 127, 129–30, 132–33, 136–37, 140–42; for Presidio plan, 111
Pelosi, Ronald, 60
People for Open Space, 55
People for the Golden Gate National Recreation Area (PFGGNRA), 57, 88, 179, 197, 231 n. 24
People for the Parks, 179
People for the Presidio, 179
pet cemetery, 105, 128, 240 n. 53
Petulla, Joseph, 151
Pinchot, Gifford, 39
Pittsburgh (Pennsylvania), 199
planning. *See* conversion planning; Presidio Planning and Transition Team
planting programs, 30–35, 105–6
Point Reyes National Seashore, 55, 57
politics: of military base closures, 78–85, 87–88; of parks, 145–46, 150–51, 167–70, 182–85, 194; of transition, 114–17
Portolá, Gaspar de, 7

Preservation of Historic Sites Act, 157
Presidio Concepts Workbook, The, 98
Presidio Council, 91–92
Presidio Forest, 3–4, 62, 106
Presidio Forum Displays, The, 98
Presidio Gate, 36 fig. 3.7
Presidio Hill, 105–6
Presidio Partnership, 119–22, 122 table 7.1 (*see also* legislation; Presidio Trust)
Presidio Plan, 111, 114–17, 114 fig. 7.1, 181–82, 199–200
Presidio Planning and Transition Team, 92–93, 98–100, 108
Presidio Planning Guidelines, 96 table 6.1, 98
Presidio Railroad, 30–31
Presidio Society, 52
Presidio Trust (*see also* Presidio Partnership): accountability to Congress, 142, 189; community mobilization for, 136–37; legislation for, 140–43, 144–45 fig. 7.4, 182–85, 242–43 n. 88, 244 n. 97, 244 n. 102, 244 n. 103; naming of, 125–27; as public-private partnership, 186
Presidio Update newsletter, 93
Presidio Wall, 34–35, 35 fig. 3.6
presidios, 9–12
private business ventures, at Presidio, 88–89, 126–27, 239–40 n. 49, 239 n. 46
pro bono support, 92, 230–31 n. 22
public awareness, 147–49, 199–200
public-private partnerships, 119–22, 186, 235–36 n. 21, 236 n. 24, 238 n. 37
public recreation areas, 159 table 8.2
pueblos, 9–12

Reagan, Ronald, 73, 74–75, 81–82, 86, 226 n. 56, 235–36 n. 21
real estate values, 88
recreation areas, 105, 158–62, 164 fig. 8.2, 196
Reilly, William, 124–25, 133, 137, 143, 176, 184
Relph, Edward, 64–65
Renn, Gregg, 115
Rettie, Dwight, 161, 163, 175
Reveille newsletter, 93
Ridenour, James, 170, 173–76, 177
Rockefeller, Nelson, 51
Roosevelt, Franklin D., 48, 157, 169
Rosenblatt, Toby, 91, 124, 143
Roth, William, 80
Rothman, Hal, 149
Runte, Alfred, 149, 152–53, 185

St. Francis of Assisi, 8
Sal, Hermengildo, 14
sale proposal, 131–36
San Andreas fault, 40
San Francisco Bay, 12 fig. 2.2, 17, 18
San Francisco Bay Association, 55
San Francisco (California): building boom, 53; city planning, 35–39, 92–93; growth of, 21–22, 25–26, 53–54; high-rise development, 53–54; Master Plan of, 231 n. 25; in post-Gold Rush era, 28–30; and Presidio Trust legislation, 185; Presidio's linkage to, 179, 228–29 n. 8; reaction to Grand Vision, 111
San Francisco National Cemetery, 105
San Francisco Peninsula, 8–9
San Francisco State University, 96
Schmitz, Eugene, 40
Schwab, Michael, 139 fig. 7.3, 147, 148 fig. 8.1
Seattle (Washington), 198
sequent occupance, 65
Serra, Junipero, 7–8
Seth, Vikram, 64
shipyards, 50
Sierra Club, 55–56, 137, 199, 200
Silverstein, Rich, 148
Sixth Army, 52, 62, 117, 138, 141, 241 n. 72, 243–44 n. 93
social theory, 190
Sola, Pablo Vincente de, 15
Spain: colonization approach, 9–12; early exploration, 7–8; hostility with Great Britain, 15; occupation at Presidio, 13–17
Spanish-American War, 38
Spanish Revival architecture, 48, 220 n. 5
special interest groups, 89
spending priorities, 185
Starr, Kevin, 21, 89–90
stewardship, 98, 99 table 6.2
Stickley, John George, 220 n. 5
Stickley, Leopold, 220 n. 5
Stinson Beach, 57
Strategic Defense Initiative (SDI), 73
sustainability, 98, 99 table 6.2
Sutro, Adolph, 33

Thoreau, Henry David, 27, 162
Tocqueville, Alexis de, 192
Toronto (Ontario), 199
transcendentalist movement, 151–52
transition phase, 92–93, 98–100, 108, 114–17

Treasure Island, 50
tree-planting programs, 33, 34 fig. 3.5
Truman, Harry S., 51, 221 n. 30
Twin Peaks, 47

United Nations, 51
University of California, San Francisco, 112, 141, 237–38 n. 34
urban national recreation areas, 160–61, 173
urban open space, 198–99
urban parks, 189–90, 197–98
urban recreation areas, 189–90
urban reforms, 26–28
U.S. Army: budget issues, 135, 242 n. 80; headquarters relocation, 30; jurisdiction challenges, 59–60; lease of land to San Francisco, 45; negotiation with Park Service, 117–19; Presidio occupation by, 22–26; in transfer ceremony, 138–40

Vail Agenda, 170, 175, 181
Vallejo, Mariano Guadalupe, 17, 18
Vancouver, George, 14–15
Vento, Bruce, 117, 129, 138, 235 n. 11
Ventresca, Joel, 126, 127, 239–40 n. 49
Veterans Administration, 105

Visions Kit, 98
Vizcaíno, Sebastian, 7

water pollution, 54
Wayburn, Edgar, 57, 231 n. 24
Weinberger, Caspar, 79
Wellman, Douglas, 149
westward expansion, 18–19
wetland reclamation, 104
wilderness, perceptions of, 152, 191–93, 245 n. 10
Wilderness Act, 169, 191–92
Wilson, Pete, 136
Wilson, Woodrow, 156
Winks, Robin, 173–74
Wirth, Conrad, 149
Work, Hubert, 153
Works Progress Administration (WPA), 48, 157
World War I, 45–46
World War II, 50–51
Wright, Frank Lloyd, 220 n. 5

Yellowstone National Park, 151, 152, 162–63, 165, 166, 194, 252–53 n. 43
Yerba Buena Pueblo, 12 fig. 2.2, 21–22, 218 n. 4
Yosemite National Park, 151, 152, 161, 162–63, 165, 194, 245 n. 8